The Bible & Ancient Science

About the Cover

The image comes from Martin Luther's 1534 translation of the Bible, and it appears across from the first chapter of Holy Scripture—Genesis 1 and the creation of the world. Scientists at that time believed that the earth was spherical, immovable, and located in the center of the entire universe. This ancient understanding of astronomy is known as "geocentrism" (Greek word *gē* means "earth"). Heaven included a solid outer sphere termed the "firmament." One geocentric theory claimed that the sun, moon, and stars were placed in the firmament, and its daily rotation caused day and night on earth.

In his 1536 *Lectures on Genesis*, Luther attempted to align the Bible with this ancient understanding of the structure and operation of the world. This approach to interpreting Scripture is called "scientific concordism" (or simply "concordism"). In commenting on the second day of creation in Genesis 1:6-8, Luther argues that God made the firmament so that "it should extend itself outward in the manner of a sphere." He adds, "Scripture . . . simply says that the moon, the sun, and the stars were placed in the firmament of the heaven . . . The bodies of the stars, like that of the sun, are round, and they are fastened to the firmament like globes of fire."

Martin Luther demonstrates the problem with scientific concordism and attempts to align Scripture with the science-of-the-day. As science advances, new facts about the natural world are discovered, and concordist interpretations are then proven to be incorrect. For example, no one today accepts Luther's ancient astronomy and his geocentric view of the universe. Moreover, should any Christian cling to scientific concordism and make it an essential component of their faith, new scientific discoveries may damage their belief in both God and the Bible. In this book, we will examine a way to move beyond concordism that honors Scripture as the Holy Spirit-inspired Word of God.

The Bible & Ancient Science
Principles of Interpretation

Denis O. Lamoureux
PhD Theology & PhD Biology

McGahan

The Bible & Ancient Science: Principles of Interpretation
Copyright © 2020 by Denis O. Lamoureux

All rights reserved. No part of this publication may be reproduced, stored in a retrieval system, or transmitted in any form or by any means—electronic, mechanical, photocopy, recording, or any other—except for brief quotations in printed reviews, without the prior permission of the publisher.

Scripture quotations marked NRSV are taken from the New Revised Standard Version Bible, copyright © 1989 National Council of the Churches of Christ in the United States of America. Used by permission. All rights reserved worldwide.

Scripture quotations marked (NIV) are taken from the Holy Bible, New International Version®, NIV®. Copyright © 1973, 1978, 1984, 2011 by Biblica, Inc.™ Used by permission of Zondervan. All rights reserved worldwide. www.zondervan.com. The "NIV" and "New International Version" are trademarks registered in the United States Patent and Trademark Office by Biblica, Inc.™

Scripture quotations marked (NASB) are taken from the New American Standard Bible ® (NASB), Copyright © 1960, 1962, 1963, 1968, 1971, 1972, 1973, 1975, 1977, 1995 by The Lockman Foundation. Used by permission. www.Lockman.org.

McGahan Publishing House
Tullahoma, Tennessee 37388, U.S.A.
www.mphbooks.com
Requests for information should be sent to:
info@mphbooks.com

ISBN 978-1-951252-05-2

Library of Congress Control Number: 2020939811

Contents

INTRODUCTION
Is the Bible a Book of Science? 11

HERMENEUTICAL PRINCIPLE 1
Literalism 15

HERMENEUTICAL PRINCIPLE 2
Literary Genre 20

HERMENEUTICAL PRINCIPLE 3
Scientific Concordism & Spiritual Correspondence 26

HERMENEUTICAL PRINCIPLE 4
Eisegesis vs. Exegesis 32

HERMENEUTICAL PRINCIPLE 5
Ancient & Modern Phenomenological Perspectives 39

HERMENEUTICAL PRINCIPLE 6
The Message-Incident Principle 45

HERMENEUTICAL PRINCIPLE 7
Biblical Accommodation 53

HERMENEUTICAL PRINCIPLE 8
Authorial Intentionality: Divine & Human 59

HERMENEUTICAL PRINCIPLE 9
Biblical Sufficiency & Human Proficiency 64

HERMENEUTICAL PRINCIPLE 10
Modern Science & Paraphrase Biblical Translation 70

HERMENEUTICAL PRINCIPLE 11
Textual Criticism 75

HERMENEUTICAL PRINCIPLE 12
Implicit Scientific Concepts 82

HERMENEUTICAL PRINCIPLE 13
Scope of Cognitive Competence 90

HERMENEUTICAL PRINCIPLE 14
Historical Criticism 98

HERMENEUTICAL PRINCIPLE 15
The 3-Tier Universe: Ancient Geography 108

HERMENEUTICAL PRINCIPLE 16
The 3-Tier Universe: Ancient Astronomy 123

HERMENEUTICAL PRINCIPLE 17
The Accommodation of God's Creative Action in Origins 137

HERMENEUTICAL PRINCIPLE 18
De Novo Creation of Living Organisms: Ancient Biology 146

HERMENEUTICAL PRINCIPLE 19
Does Conservative Christianity Require Scientific Concordism? ... 155

HERMENEUTICAL PRINCIPLE 20
Literary Criticism 162

HERMENEUTICAL PRINCIPLE 21
Source Criticism 178

HERMENEUTICAL PRINCIPLE 22
Biblical Inerrancy: Toward an Incarnational Approach 185

CONCLUSION
Beyond Scientific Concordism 191

APPENDIX 1
Christian Positions on the Origin of the Universe & Life 198

APPENDIX 2
The "Waters Above" & Scientific Concordism 199

APPENDIX 3
Do Isaiah 40:22 & Job 26:7 Refer to a Spherical Earth? 203

NOTES 207

Acknowledgements

One of the greatest blessings in my life has been teaching a university course on the relationship between science and religion. Over the years, many of my students have told me that discovering principles of biblical interpretation was the key that led them to a peaceful relationship between modern science and their Christian faith. In fact, over eighty percent of the class in my evaluations state that learning how to interpret creation accounts and statements about nature in Scripture was "the most valuable aspect of the course." It is thanks to the encouragement of my students that has inspired me to write this book.

I am so grateful to many people who have assisted me in this project. My teaching assistant Anna-Lisa Ptolemy worked tirelessly and meticulously on editing numerous versions of the manuscript. It was a wonderful experience to collaborate with Caleb Poston, the CEO of McGahan Publishing House. Thank you to Duane Cross for his marketing expertise. Many thanks to Andrea Dmytrash and Kenneth Kully for their artwork. And I am forever grateful for the support of President Terence Kersch and Dean Shawn Flynn at St. Joseph's College in the University of Alberta.

Others who have contributed to my work include Angela Anderson-Konrad, Ed Babinski, Chris Barrigar, Lyn Berg, Reily Cross, Mona-Lee Feehan, Keith Furman, Sy Garte, Brian Glubish, Wendell Grout, Loren Haarsma, David Haitel, Reagan Haitel, Douglas Jacoby, Donald Johnson, Bob Lamoureux, Jack Maze, Don McLeod, Sara McKeon, Isabel Ptolemy, Don Robinson, Jim Ruark, Anita and Paul Seely, Jennifer Swainson, Kevin Tam, Madison Trammel, and Loren Wilkinson.

"*The Bible is a book of science!* The Bible does contain all the basic principles upon which true science is built."
    Dr. Henry Morris
    Father of Modern Young Earth Creationism

"*The Bible is not a book of science.* The Bible is a book of redemption."
    Rev. Billy Graham
    Greatest Preacher of the 20th Century

INTRODUCTION

Is the Bible a Book of Science?

The Bible is a precious gift from God. It reveals who our Creator is and who we are. Scripture sets down the foundational beliefs of the Christian faith. The Word of God affirms the creation of the universe and living organisms, the sinfulness of all men and women, the offer to restore our relationship with the Lord through the sacrifice of Jesus on the Cross, and the hope of eternal life. The Bible is an everlasting spring of spiritual living waters for our thirsty souls. Through the gracious guidance of the Holy Spirit, it assures and encourages, at times challenges and admonishes, and also equips Christians for a life of good works. Most importantly, the central purpose of the Word of God is to reveal our Father and his love for each and every one of us.

I personally know the life-changing power of the Bible. By the grace of God and in answer to the prayers of my mother, I read the Gospel of John while serving in the army as a United Nations Peacekeeper on the island of Cyprus over forty years ago. Through this marvellous biblical book, the Holy Spirit convicted me of my shameful lifestyle, and I accepted the forgiveness of my sins through Jesus. By reading Scripture I grew in the faith and came to enjoy the peace and freedom of walking with the Lord. Indeed, a military peacekeeper met the Prince of Peace and was born-again!

Not long after my conversion to Christianity, I started to think about the topic of origins. My education at a secular university was in biology and dentistry, and I had been thoroughly indoctrinated into an atheistic view of evolution. To move forward as a Christian, I needed to deal with the issue of origins. I was surprised to discover that there were three different Christian views on how God created the world. Most of the people in my church accepted young earth creation. They believed that the entire cosmos and all plants and animals were made in six literal days about six thousand years ago. I quickly and firmly embraced this position because it aligned with how I read and interpreted the biblical account of creation in Genesis 1.

A geologist in my church then introduced me to a second understanding of origins. Progressive creation asserts the universe is very old and that God miraculously created various living organisms at different times during the 4.6-billion-year history of the earth. Also known as "old earth creation" and "day-age creation," it interprets the six days of creation in Genesis 1 as six periods of time that are millions of years long. Because of my literal way of reading Scripture at that time, I found this view completely unacceptable and unbiblical.

Finally, Christians in my church warned me about theistic evolution, which today is called "evolutionary creation." This position claims that God used evolution to create the universe and life, and that the purpose of the Bible is to reveal spiritual truths and not scientific facts about origins. Evolutionary creationists believe that in order to reveal these messages of faith to ancient people, the Holy Spirit allowed the biblical writers to use their ancient understanding of origins. In other words, the Genesis 1 creation account features what could be termed an "ancient science." Every Christian I knew despised this view of origins. Many of us assumed that Satan had concocted the lie of evolution, and only so-called "liberal" Christians were deceived by this false science. Appendix 1 summarizes the three Christian positions on origins.

Today the topic of origins continues to be a critical issue facing the church. As theologian Scot McKnight observes, "The number one reason young Christians leave the faith is the conflict between science and faith, and that conflict can be narrowed to the conflict between evolutionary theory and human origins as traditionally read in Genesis 1-2."[1] McKnight also adds, "[T]he number one reason non-Christians find the Christian faith untrustworthy is the issue of the Bible and science."[2] I have personally experienced this conflict in my life. As a freshman college student, I rejected Scripture and Christianity because of evolutionary science; later as a born-again Christian, I rejected evolution and modern science because of my literal interpretation of Genesis 1 and 2.

This book deals with one primary question: Is the Bible a book of science? As the epigraphs found across from the first page of this chapter reveal, important Christian leaders have offered completely different answers to this question. On the one hand, the famous young earth creationist Dr. Henry Morris claims, *"The Bible is a book of science!* The Bible does contain all the basic principles upon which true science is built."[3] On the other hand, the renowned preacher of the gospel Rev. Billy Graham asserts, *"The Bible is not a book of science.* The Bible is a book of redemption."[4]

In order to answer this question of whether Scripture is a book of science, we need to deal directly with biblical interpretation. This topic is known as "hermeneutics." Simply defined, it refers to the principles used for interpreting a piece of literature. For me, biblical hermeneutics is one of the most fascinating subjects in theology, and it is like the practice of science in that it is a godly activity ordained by our Creator. The famous seventeenth century astronomer Johannes Kepler once said, "Those laws [of nature] are within the grasp of the human mind; God wanted us to recognize them by creating us after his own image so that we could share in his own thoughts."[5] In a similar way, I view the practice of hermeneutics as sharing in the thoughts of the Holy Spirit during the process of inspiring the biblical authors to create the Word of God.

This book focusses on the interpretation of passages in Scripture that deal with the natural world and the biblical creation accounts in Genesis 1 and 2. Whether we are completely aware of it or not, we all have hermeneutical notions and assumptions, because we all read and interpret the Bible. Yet today many Christians simply read Scripture without fully appreciating their interpretive methods. My hope and prayer for this book is that it helps you to be more mindful of your hermeneutical principles. In this way, you will make informed decisions about the interpretation of biblical passages referring to nature and origins and become better interpreters of the Word of God.

HERMENEUTICAL PRINCIPLE 1

Literalism

Many Christians today assume that reading the Bible literally is the correct and most faithful way to interpret the Word of God. There was a time when I believed that *true* Christians demonstrated their commitment to the Lord by believing in the literal interpretation of Scripture, in particular the creation accounts in Genesis 1 and 2. To be sure, there are a lot of literal passages in the Bible, such as the literal bodily resurrection of Jesus from the grave after his literal physical death on the Cross. However, there are also several passages in Scripture that cannot be read literally. Let's look at a few of these.

Isaiah 55:12 states, "You will go out in joy and be led forth in peace; the mountains and hills will burst forth in song before you, and all the trees of the field will clap their hands." Of course, we all know that mountains and hills cannot sing, and trees do not have hands. To offer another example, Psalm 91:4 records, "He [God] will cover you with his feathers, and under his wings you will find refuge." I am doubtful that any Christian reads this passage literally and believes that God is some sort of cosmic bird!

In Matthew 5:28-29, Jesus commands, "I tell you that anyone who looks at a woman lustfully has already committed adultery with her in his heart. If your right eye causes you to stumble [sin], gouge it out and throw it away. It is better for you to lose one part of your body than for your whole body to be thrown into hell." I have been in many churches throughout America and rarely have I seen men without eyes. Does this mean that Christian men never have lustful thoughts when looking at a woman? Or are they being unfaithful by disregarding the Lord's command to pluck their eye out when they do lust? And since this passage is directed at men, does it mean women never lust? I suspect the ladies are probably grinning right about now!

As these three passages reveal, it is impossible to read the Bible literally 100% of the time. In fact, Jesus did not speak literally on every occasion. For example, in John 16:25 he openly states, "Though I have been speaking figuratively, a time is coming when I will no longer use this kind of language but will tell you plainly about my Father." In verse 29, the Lord's disciples noted this change in his teaching style. "Now you are speaking clearly and without figures of speech."

As everyone knows, a figure of speech employs words in an imaginative way and is not intended to be understood literally. One type of figure of speech is hyperbole. This is the deliberate use of exaggeration. It is an effective way to capture the attention of people and Jesus' teaching on lust is a good example of this approach. Personification is another figure of speech. It attributes human characteristics to inanimate objects, and also uses things to represent people. Jesus employed this technique in John 10:9. "I am the gate; whoever enters through me will be saved." Similarly, in John 15:5 the Lord told his disciples, "I am the true vine; and you are the branches. If you remain in me and I in you, you will bear much fruit; apart from me you can do nothing." Obviously, Jesus is not literally a gate or a vine, and we are not literally the branches of a vine.

In addition, every Christian is aware that Jesus taught using parables. These are made-up stories with a spiritual truth. In fact, about one third of the Lord's teaching was through parables. One of his best-known is the parable of the Good Samaritan in Luke 10:30-37.

> A man was going down from Jerusalem to Jericho, when he was attacked by robbers. They stripped him of his clothes, beat him and went away, leaving him half dead. A priest happened to be going down the same road, and when he saw the man, he passed by on the other side. So too, a Levite, when he came to the place and saw him, passed by on the other side. But a Samaritan, as he traveled, came to where the man was; and when he saw him,

> he took pity on him. He went to him and bandaged his wounds, pouring on oil and wine. Then he put the man on his own donkey, brought him to an inn and took care of him. The next day he took out two silver coins and gave them to the innkeeper. "Look after him," he said, "and when I return, I will reimburse you for any extra expense you may have."

Jesus then asked, "Which of these three do you think was a neighbor to the man who fell into the hands of robbers?" Someone answered, "The one who had mercy on him." And the Lord said to this person, "Go and do likewise."

Did the events described in the parable of the Good Samaritan literally happen? No. They are part of a story that Jesus made up in order to reveal a spiritual truth about having mercy on anyone who needs help. The power of the Lord's message in this parable is that the story can be updated for any generation, including ours today. For example, we could re-write this parable using an account about road rage:

> A man was going down from New York City to Washington DC, when he was driven off the highway. The other driver and his passenger got out of their car and beat up the man, leaving him half dead. A Protestant pastor happened to be going down the same highway, and when he saw the man in the ditch, he passed by him. So too, a Catholic priest, when he came to the place and saw him, passed by him. But a non-Christian, as he traveled, came to where the man was; and when he saw him, he took pity on him. He went to him and bandaged his wounds, using anesthetic and antiseptic ointments. Then he put the man in his car, brought him to a hotel and took care of him. The next day he took out five hundred dollars and gave it to the hotel manager. "Look after him," he said, "and when I return, I will reimburse you for any extra expense you may have."

The events in this story never happened. I made them up. But I believe that this modern-day account of road rage captures the central message that Jesus taught in his parable of the Good Samaritan. The updated story is a vessel that faithfully transports the Lord's spiritual lesson that we need to be merciful to all our neighbors, no matter who they are or where they come from. In this way, *non-literal accounts can deliver life-changing messages of faith*.

To conclude, the use of parables and figures of speech by Jesus is proof that the Bible includes passages that cannot be read literally. Moreover, the Lord's made-up stories and figurative language are solid evidence that literal passages are not more important or holier than non-literal passages. To help my students remember this first interpretive principle, I call it Hermeneutical Commandment #1: Thou shalt not believe that the Bible is 100% literal!

Of course, the question you may be asking right now is, when should we read Scripture literally? There is no fast and easy answer other than to say that if we learn and practice basic hermeneutical principles, then we will become more skillful in determining if a biblical passage is intended to be read in a literal way. By the end of this book, I guarantee that you will be more equipped to answer this question.

It is necessary to emphasize that everyone who reads Scripture must make a decision on whether or not to read a biblical passage literally. And everyone makes this interpretive choice, whether or not they are fully aware of it. The purpose of this book is to encourage you to think about your hermeneutical decisions so that you can make informed choices. In particular, we will focus on biblical passages dealing with the natural world and attempt to determine if they should be read literally or non-literally.

Biblical Creation Accounts

This is probably one of the most crucial questions for Christians interested in the topic of origins: Is a literal reading of the accounts of creation in Scripture the correct interpretation? Many Christians throughout history have understood Genesis 1 and 2 to be a word-for-word account of how God actually created the universe and living organisms.

For example, the famous sixteenth century theologian Martin Luther believed that the writer of Genesis 1 "spoke in the literal sense, not allegorically or figuratively; that is, that the world, with all its creatures, was created within six days, as the words read."[1] A survey of adults in the United States reveals a similar hermeneutical approach. They were asked about the creation account in Genesis 1, "Do you think that's literally true, meaning that it happened that way word-for-word; or do you think it's meant as a lesson, but not to be taken literally?"[2] Sixty-one percent of Americans believe that "the creation story in which the world was created in six days" was "literally true." And eighty-seven percent of evangelical (born-again) Christians think that Genesis 1 is a literal account of how God actually created the world.

It is important to point out that Jesus himself referred to the biblical creation accounts in Genesis 1 and 2. In responding to his critics with regard to the issue of divorce, Matthew 19:4-5 records, "'Haven't you read,' he [Jesus] replied, 'that at the beginning the Creator 'made them male and female' [Gen. 1:27],' and said, 'For this reason a man will leave his father and mother and be united to his wife, and the two will become one flesh?' [Gen. 2:24]."

In this passage Jesus appeals directly to the creation of humans in Genesis 1, and in particular to Adam and Eve in Genesis 2. Therefore, Christians who claim that the opening chapters of Genesis should not be read literally need to offer convincing reasons why these accounts are not a word-for-word record of actual events in the creation of the first humans. In proceeding through this book, we will explore various explanations for non-literal interpretations of Genesis 1 and 2. And we will return to Matthew 19:4-5 and attempt to offer a persuasive approach to reading this passage.

HERMENEUTICAL PRINCIPLE 2

Literary Genre

Many Christians are aware that the Bible has a wide variety of different types of literature, known as "literary genres." To mention a few of these, Scripture includes poetry, hymns, proverbs, prophecies, sermons, stories, parables, allegories, genealogies, narratives, historical reports, personal letters, and gospels with eyewitness accounts of real events. As Hebrews 1:1 states, "In the past God spoke to our ancestors through the prophets at many times and *in various ways*" (my italics). Sometimes the literary genre of a passage in Scripture is unfamiliar to modern readers because it is an ancient type of literature that is rarely if ever used today.

For example, in the previous hermeneutical principle on literalism, we examined a passage where Jesus tells his listeners to tear out their eyes should they look at someone lustfully (Matt. 5:27-28). These verses appear in a section (Matt. 5:21-48) within Jesus' well-known Sermon on the Mount (Matt. 5:1-7:29). And like the command to pluck out our eyes whenever we lust, the Lord makes a number of other shocking statements.

Jesus asserts, "Anyone who is angry with a brother or sister will be subject to judgment" and "anyone who says 'You fool!' will be in danger of the fire of hell" (v. 22). It strikes me as extremely severe that our eternal judgment and destiny could be determined by simply being angry with someone or just calling them a fool.

The Lord also tells his listeners, "If your right hand causes you to stumble [sin], cut it off and throw it away. It is better for you to lose one part of your body than for your whole body to be thrown into hell" (v. 30). Most Christians would acknowledge that we are sinners (1 Jn. 1:8), but we do not follow this command literally, because our churches are not filled with handless men and women.

Jesus then commands, "Do not resist an evil person. If anyone slaps you on the right cheek, turn to them the other also" (v. 39). I am sure you will agree that no society can function by allowing evil people to do whatever they want without being opposed. In fact, Christians do not believe that a police officer is acting against God's will if he or she stops someone from committing a crime. Instead, we are grateful for their courage and service.

The Lord adds, "If anyone wants to sue you and take your shirt, hand over your coat as well" (v. 40). I doubt that any Christian being sued by someone would feel compelled to follow this command. Surely, Christian lawyers do not use this principle in their practice of the law.

In addition, Jesus orders, "Give to the one who asks you, and do not turn away from the one who wants to borrow from you" (v. 42). If we followed this teaching, the possibility arises that all our possessions could be given away. Undoubtedly, there are people who would take advantage of us. And if Christian bankers gave a loan to every person who wanted one, including those without the ability to repay it, their bank would soon fail and close its doors.

And in a well-known verse, the Lord teaches, "Love your enemies and pray for those who persecute you" (v. 43). Does this mean that during World War II Christians in the American, Canadian, and British militaries were not following God's will when they launched attacks on Hitler and the Nazis?

It seems obvious that we cannot read these six verses above and then follow Jesus' commands literally. I am not aware of any Christian who does so. In fact, a serious conflict between the Lord's words and his actions would arise in Matthew 5:22 with regard to being angry and calling someone a fool. Mark 3:5 records that Jesus looked at some people "in anger," and Matthew 23:17 states that he called teachers and Pharisees "blind fools." These passages are more evidence that it is impossible to read the Bible literally 100% of the time.

The key to interpreting Matthew 5:21-48 in the Sermon on the Mount begins with determining the literary genre of this passage. What type of literature is this? First, it is part of a sermon that is directed at an audience of first century Jewish men and women. The introductory clauses "You have heard that it was said . . ." and "It has been said . . ." refer to passages in the Old Testament and religious practices in their community.

Second, it is quite clear that Jesus is employing hyperbole. These are deliberate overstatements and intended exaggerations. Therefore, they are not literal commands that we must obey word-for-word. This literary technique is quite effective in capturing the attention of the Lord's audience, as well as our generation today. Are there any Christians who can't remember Matthew 5:28-29 and the command to pluck out their eye for looking at someone lustfully?

In considering these two main characteristics, I have yet to discover a modern literary genre category that compares to the type of literature found in Matthew 5:21-48 of the Sermon on the Mount.[1] Nor do I know of anyone who uses this genre today as a teaching tool. For example, are there pastors who preach sermons using hyperboles like those of Jesus? I have never heard such a sermon in my forty years of attending church. This is sometimes the challenge of biblical interpretation. There are ancient literary genres that are quite unfamiliar to our modern generation, like this section in the Sermon on the Mount. Yet, Matthew 5:21-48 certainly speaks to us and challenges us. In particular, the Holy Spirit works through Jesus' hyperboles to convict us of our thoughts, attitudes, and sinfulness.

One of the most important decisions in biblical hermeneutics is to identify the type of literature being used in Scripture. To be more precise, the literary genre of a passage *dictates* how it is to be interpreted. If we misidentify the genre of a passage, then we will misinterpret the passage. For instance, if we fail to recognize that Jesus is using hyperboles in Matthew 5:21-48 and assume that he is making literal commands like the Ten Commandments (Exod. 20:1-17), then there would be a lot of sightless and handless Christians! Determining the literary genre of a passage in Scripture is a key to its interpretation.

Biblical Creation Accounts

One of the greatest hermeneutical challenges that Christians have faced throughout the ages is determining the literary genre of the creation accounts in the Book of Genesis. It is worth noting that some church leaders did not read these passages literally and viewed them as figurative and allegorical stories.[2] However, history reveals that a majority of Christians have consistently believed that the opening chapters of Scripture are a word-for-word historical account of actual events in the past.

Let's consider the views of the most important young earth creationist in the world today. Ken Ham is the president of both Answers in Genesis and the Creation Museum in Kentucky. In his book *The Lie: Evolution. Genesis—The Key to Defending Your Faith*, he writes:

> Many Christians fail to realize that the events of Genesis are literal, are historical (particularly Genesis 1-11), and are foundational to all Christian doctrine. All biblical doctrines of theology, directly or indirectly, ultimately have their basis in the Book of Genesis. Therefore, a believing understanding of the Book of Genesis is a prerequisite to an understanding of God and His meaning to man. If Genesis is only a myth or allegory, then Christian doctrines have no foundation.[3]

Ken Ham raises several significant issues. As we saw in the previous hermeneutical principle on literalism, Jesus made up a story in the parable of the Good Samaritan in order to reveal a spiritual truth about showing mercy to our neighbors. Even though the events in this parable are not literal or historical, I do not think for a moment that the Lord's message about mercy has "no foundation."

I certainly agree with Ken Ham that the spiritual truths in the opening chapters of the Book of Genesis are "foundational to all Christian doctrine." For example, the God of the Bible is the Creator of the world (Gen. 1:1), humans are created in the Image of God (Gen. 1:26-27), men and women are sinful (Gen. 3:6), and God judges people for their sins (Gen. 3:16-19). However, if the Holy Spirit decided to reveal these spiritual messages by inspiring the biblical writers to use a literary genre similar to a parable or an allegory, then that is God's decision. Our goal as interpreters of Scripture is to recognize and respect the type of literature employed by the Lord, and not force into Scripture our assumptions about how the Holy Spirit is supposed to have revealed in the Bible.

To explain what I mean, consider Genesis 2-3 and the Garden of Eden account. Genesis 2:9 states that the Lord God created two mystical trees in the middle of a garden paradise—the tree of life and the tree of the knowledge of good and evil. In Genesis 3:1-5 there is a fast-talking snake who tempts the woman into breaking God's command of not eating from the second tree. If you were to discover a document that mentioned trees that could impart eternal life and moral knowledge along with a talking animal, would you immediately claim that this document is a historical account about literal and actual events? Or would you say that it seems to be an allegory or made-up story like a parable to teach a spiritual or moral lesson?

Before we go any further, I need to make an important comment about historical events that are recorded in the Bible. *I firmly believe that real history in Scripture begins roughly around Genesis 12 with Abraham and his family.*

So, from my perspective, was Abraham a real person? Yes. Did Moses exist and did God reveal the Ten Commandments to him on Mount Sinai? Yes. Did Israel attack and defeat the Canaanites during the mid-thirteenth century BC? Yes. Was there a King David in the tenth century BC as mentioned in 2 Samuel 2? Yes. Was Israel attacked by the Assyrians in the late eighth century BC as stated in 2 Kings 18? Yes. Were the Jews deported to Babylon in the sixth century BC as recorded in Jeremiah 52? Yes. In fact, the science of biblical archeology and ancient historical records from countries surrounding the nation of Israel confirm that many events in Scripture actually happened.[4]

LITERARY GENRE

But more importantly, was there truly a man named Jesus in the first century AD? Yes. Did individuals like King Herod mentioned in Matthew 2 and Pontius Pilate in Matthew 27 really exist? Yes. And are the gospels based on eyewitness accounts of actual historical events, including the Lord's teaching and miracles, and especially his bodily resurrection from the grave three days after his physical death on the Cross? *Absolutely yes!* I thoroughly believe that the Bible records real events about the history of Jesus and his life.[5]

Christians are firmly united in believing in a literal and historical Jesus, and that he is our Lord and Savior. But many of us debate intensely over the interpretation of the biblical accounts of creation. As we noted in the Introduction, there are three different Christian positions on the origin of the universe and life—young earth creation, progressive creation, and evolutionary creation (Appendix 1). It is obvious that they cannot all be right. In fact, two of these three positions are wrong. So, which view of origins is correct?

The answer to this question begins by first determining the literary genre of the creation accounts in Scripture. Remember, the literary genre of a biblical passage *dictates* how it is to be interpreted. And it may be that the accounts of origins in Genesis 1 and 2 feature an ancient type of literature that is not commonly found today, just like the unfamiliar literary genre featuring hyperbole that Jesus used in Matthew 5:21-48 of his Sermon on the Mount. Are you open to the possibility of re-thinking the literal interpretation of the opening chapters of the Word of God?

HERMENEUTICAL PRINCIPLE 3

Scientific Concordism & Spiritual Correspondence

Is the Bible a book of science? Important Christian leaders answer this question in radically different ways. Some claim that statements in Scripture about the physical world align with the facts of nature discovered by modern science. Others believe that the purpose of the Bible is not to reveal scientific facts, but rather spiritual truths about God, his creation, and us. Let's revisit two faithful Christians that we have briefly mentioned in the Introduction and examine their very different approaches to the relationship between Scripture and science.

Henry Morris founded the Institute for Creation Research and he was the most important young earth creationist of the twentieth century. He firmly declares without any hesitation:

> The real truth of the matter is that the Bible indeed is verbally inspired and literally true throughout. Whenever it deals with scientific or historical matters of fact, it means exactly what is says and is completely accurate. When figures of speech are used, their meaning is always evident in context, just as in other books. There is no scientific fallacy in the Bible at all. "Science" is knowledge, and the Bible is a book of true and factual knowledge throughout, on every subject with which it deals. *The Bible is a book of science!* . . . The Bible does contain all the basic principles upon which true science is built. These principles do not pass out of fashion and have always been valid. They were recorded in Scripture long before man learned them through his scientific research.[1]

Billy Graham is without a doubt the greatest preacher of the Gospel of Jesus during the twentieth century. He approaches the issue of the Bible and science in a much different way:

> I think we have misinterpreted the Scriptures many times and we've tried to make the Scriptures say things that they weren't meant to say, and I think we have made a mistake by thinking the Bible is a scientific book. *The Bible is not a book of science.* The Bible is a book of redemption, and of course, I accept the Creation story. I believe that God did create the universe. I believe He created man, and whether it came by an evolutionary process and at a certain point He took this person or being and made him a living soul or not, does not change the fact that God did create man . . . whichever way God did it makes no difference as to what man is and man's relationship to God.[2]

The central difference between Henry Morris and Billy Graham is the way they approach the hermeneutical concept of "scientific concordism," often referred to as simply "concordism." The word "concord" means "to be in agreement" or "in harmony." Scientific concordism is the assumption that statements about nature in the Bible align with the facts of science. Many Christians like Henry Morris assume that God revealed some basic scientific truths in Scripture well before their discovery by modern science. Concordism is often presented as proof that the Bible really is the Word of God. Only a Divine Being who is powerful and transcends time could have given modern scientific information to the ancient authors of Scripture.

It is necessary to point out that scientific concordism is a very reasonable assumption. God is the Creator of the world, and he is also the Author of the Bible. To expect an alignment or harmony between Scripture and the scientific facts of nature makes perfect sense. It is also worth noting that most Christians today are concordists (though they may not know this term). As we saw previously, sixty-one percent of adults in the United States believe Genesis 1 is "literally true, meaning that it happened that way word-for-word."[3] And eighty-seven percent of evangelical (born-again) Christians believe that Genesis 1 is a literal and historical account of how God actually created the world. Like Henry Morris, a majority of Americans are young earth creationists.

However, Billy Graham introduces us to a new way of looking at the relationship between Scripture and science. He rejects the idea that the Bible is a book of science. In fact, he states that it is "a mistake" to claim that the Word of God is a scientific text. Graham contends that such an approach is a misinterpretation of the Bible, and it forces "the Scriptures to say things that they weren't meant to say." In other words, scientific concordism is an error.

To correct this misuse of Scripture, Billy Graham underlines that the Word of God is a "book of redemption." The primary purpose of the Bible is to deal with our relationship with God and to make us realize that we have damaged it through our own sinfulness. And the Good News of the Gospel, which Graham so brilliantly preached throughout his career, is that Jesus can repair and restore this broken relationship. By dying on the Cross for our sins, the Lord has opened the way for us to have a personal and loving relationship with our Creator.

According to Billy Graham, the Bible is a book of spirituality. To introduce another hermeneutical category, "spiritual correspondence" is the belief that spiritual statements in Scripture align with spiritual reality. I prefer to use the word "correspondence" instead of "concordism," since the latter mostly appears in contexts that deal with physical reality.[4] Moreover, the term "correspondence" often has the sense of communication between individuals, like a personal letter. Therefore, spiritual correspondence refers to the belief that God personally communicates life-changing spiritual truths to men and women through the Bible.

I must point out that Henry Morris and Billy Graham were both wonderful Christians who loved the Word of God. Yet they hold completely different views on scientific concordism. This leads them to have very different positions on how God created the world. As a concordist, Morris claims, "Divine revelation from the Creator of the world [states] that He did it all in six days several thousand years ago."[5] The assumption of scientific concordism in Genesis drives Morris to accept young earth creation.

In contrast, Billy Graham believes that the Bible is not a book of science and does not reveal how God actually created the world. By not being chained to scientific concordism, he is open to the possibility that the Lord could have used an evolutionary process to create the universe and life, including men and women. According to this approach, the Bible tells us *who* created the world, while modern science reveals *how* it was created.

Biblical Creation Accounts

Figure 3-1 presents the relationship between scientific concordism and spiritual correspondence. Notably, with regard to origins in the Bible, there is an overlapping area between the statements about nature and the statements about spirituality. In my opinion, this is where the greatest challenge exists for any Christian attempting to understand the creation of the universe and life. To explain what I mean, let's consider a topic that is often discussed within the origins debate.

Genesis 1 states ten times that God created living organisms "according to their/its kinds." For example, verse 11 records, "Then God said, 'Let the land produce vegetation: seed-bearing plants and trees on

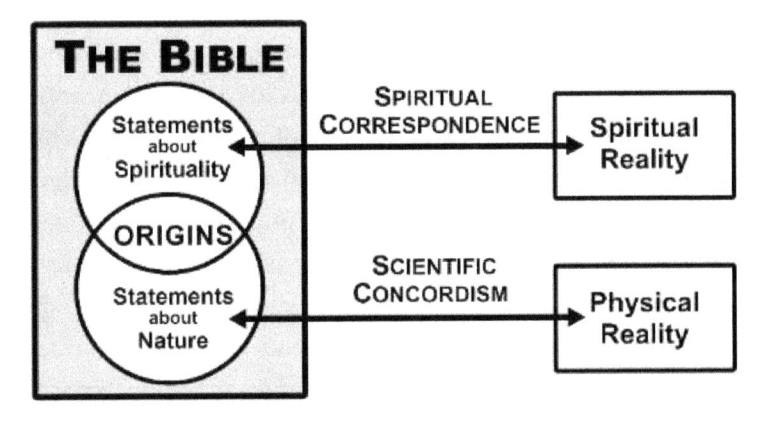

Figure 3-1. Spiritual Correspondence & Scientific Concordism

the land that bear fruit with seed in it, according to their various kinds.'" Similarly, verse 24 asserts, "And God said, 'Let the land produce living creatures according to their kinds: the livestock, the creatures that move along the ground, and the wild animals, each according to its kind.'"

Are these verses in Genesis 1 scientific statements about how God actually made different types or groups of plants and land animals? Scientific concordists claim that the phrase "according to their/its kinds" is biblical proof that the Creator did not employ evolution to create life. These anti-evolutionists argue that in creating separate "kinds" God used miraculous interventions to form each distinct type of organism individually. However, some Christians reject scientific concordism and believe that Genesis 1 does not say *how* the Lord created living creatures. Instead, they contend that this creation account reveals the foundational spiritual truth of *who* made plants, animals, and humans—the God of the Bible and Christianity.

Here are a few more thoughts I would like you to consider regarding scientific concordism and spiritual correspondence. Do the statements about the physical world in the Bible actually align with the scientific facts found in nature by modern scientists? If Scripture and modern science do not match up, will this destroy your belief that the Bible is really the Word of God? Can Christians believe that the main purpose of the biblical creation accounts is to reveal spiritual messages about the Creator, his creation, and us? Is it therefore reasonable to reject scientific concordism, but to accept spiritual correspondence? To be more precise, *does a non-concordist hermeneutic honor Holy Scripture?*

Let me introduce an idea that is not often heard in our churches and Sunday schools. There are some Christians today who believe that the reason statements about nature in Scripture do not align with the physical world is because God allowed the biblical writers to use their ancient understanding of origins in order to communicate spiritual truths. In other words, the Holy Spirit descended to the intellectual level of ancient people during the process of inspiring Scripture. According to this approach, the Bible includes their science-of-the-day, or what could be termed "ancient science."

Now some might object to the idea of ancient science and claim that people in the past did not practice science, because they did not have the precise methods and instruments of today.[6] This is a fair concern. But we need to remember that Scripture includes other ancient notions from the historical periods when the inspired authors lived. For example, ancient modes of transportation appear in Scripture, like Jesus riding on a donkey in his triumphal entry into Jerusalem (Matt. 21:1-11). Ancient cultural practices are present, such as the Lord washing the feet of his disciples before the Last Supper (Jn. 13:1-17). And ancient methods of execution are found in the Bible. Jesus died by being crucified on the Cross for our sins and salvation (Lk. 23:32-33). So, it is reasonable to expect that ancient concepts about the physical world are also part of Scripture.

To be sure, the idea that the Bible has an ancient science can be quite challenging, and even threatening to most Christians the first time they hear it. And if this is the case for you, I apologize and will ask for your patience. As we proceed through this book, we will explore numerous biblical passages so that you can make an informed decision about whether or not Scripture has an ancient understanding of nature. You will then be able to determine if scientific concordism is a feature of the Word of God.

HERMENEUTICAL PRINCIPLE 4

Eisegesis vs. Exegesis

During the first day of my college course on the relationship between science and religion, I have students read the first three verses of the Bible. "[1] In the beginning God created the heavens and the earth. [2] Now the earth was formless and empty, darkness was over the surface of the deep, and the Spirit of God was hovering over the waters. [3] And God said, 'Let there be light" and there was light.'" I then ask them to draw a diagram of the scene that they envision in Genesis 1:2.

Nearly 90% of the students sketch a water-covered *spherical* earth. Some examples of their drawings appear in Figure 4-1. When reading the word "earth," they automatically picture a globe. Yet when I ask them if they have ever heard that ancient people believed in a *flat* earth, they sheepishly say "yes" and admit that they never made the connection between the Bible and this ancient understanding of the structure of the world. I encourage them that this is one of the reasons we need to learn hermeneutical principles. They make us more aware of how to read a book written in ancient times, like the Word of God.

The interpretive error that most of my students make in picturing Genesis 1:2 as a spherical planet is known as "eisegesis." The Greek preposition *eis* means "in, into," and *ēgeomai* is the verb "to guide." Eisegesis refers to reading our own ideas or agendas *into* a passage or book. This is a common error that all of us have made at one time, and it often occurs in biblical interpretation. This is the mistake most people make by forcing the modern scientific notion of a spherical earth into the Bible when reading the word "earth" in Genesis 1:2. Many years ago, I committed this eisegetical error when I first read the Bible as a new Christian.

EISEGESIS VS. EXEGESIS

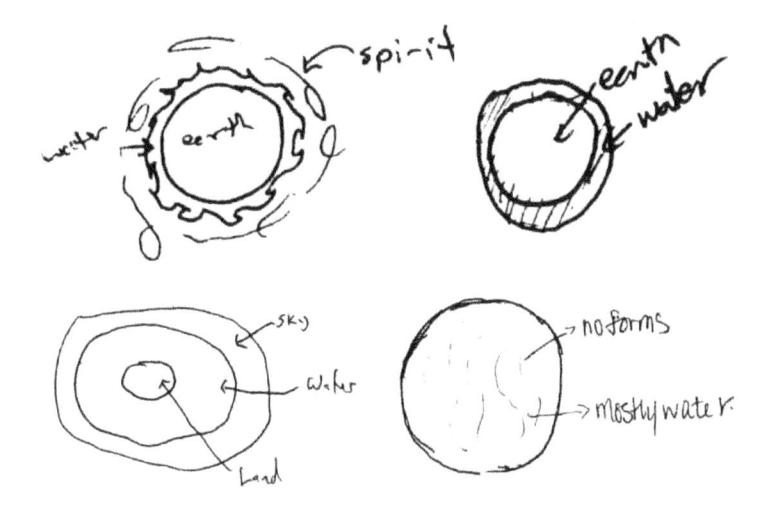

Figure 4-1. Student Diagrams of Genesis 1:2

Let's turn to another biblical passage and begin to introduce evidence that Scripture has an ancient understanding of the structure of the world. Philippians 2:5-11 is called the "Kenotic Hymn" and it is one of the most important passages in the Bible. The Greek verb *kenoō* means "to empty." This hymn reveals a foundational belief of the Christian faith—God emptied himself and became a man in the person of Jesus in order to die for our sins. The apostle Paul writes,

> [5] In your relationships with one another, have the same mindset as Christ Jesus: [6] Who, being in very nature God, did not consider equality with God something to be used to his own advantage, [7] rather, he made himself nothing [*kenoō*], by taking the very nature of a servant, being made in human likeness. [8] And being found in appearance as a man, he humbled himself by becoming obedient to death––even death on a cross! [9] Therefore God exalted him to the highest place and gave him the name that is above every name, [10] that at the name of Jesus every knee should bow, [1] in heaven and [2] on earth and [3] under the earth, [11] and every tongue confess that Jesus Christ is Lord, to the glory of God the Father.

33

THE BIBLE & ANCIENT SCIENCE

Most Christians do not notice the reference to the ancient understanding of the structure of the universe in verse 10. This is known as the "3-tier universe." According to this ancient science, the world has three levels: heaven overhead, the surface of a flat earth in the middle, and a lower region inside the earth.

For years I enjoyed singing the Kenotic Hymn during the praise and worship service in my church, but never once did I recognize this ancient understanding of the cosmos. It was only when I began to study biblical hermeneutics in seminary that I became aware of this ancient science in Scripture.*

It was also during my training in theology that I learned ancient Greek, the language used by the apostle Paul in Philippians 2:5-11. To my surprise, I discovered that the English translation "under the earth" was not completely accurate.

The actual Greek word that appears in verse 10 is *katachthoniōn*.[1] It is made up of the preposition *kata* which means "down," and the noun *chthonios* that refers to the "underworld" or "subterranean world." Therefore, a more precise translation of Philippians 2:10 would be:

At the name of Jesus every knee should bow,

> [1] in heaven
>
> [2] on earth and
>
> [3] down in the underworld.

In other words, Paul is referring to a 3-tier universe in this passage as shown in Figure 4-2.

I believe everyone will agree that the goal of reading any passage is to draw out the author's intended meaning from it. This is termed "exegesis." The Greek preposition *ek* means "out, out of," and as we have noted, *ēgeomai* is the verb "to guide." Even if we may disagree with an

---

* We will examine in more detail the many biblical passages that describe a 3-tier universe in Hermeneutical Principles 15-17.

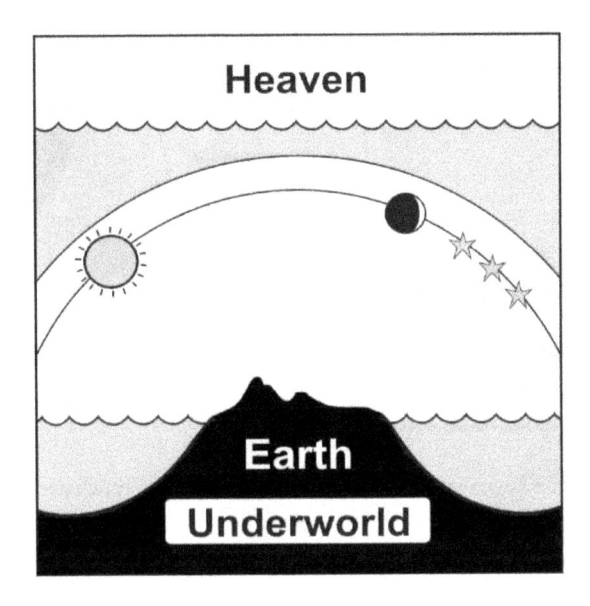

Figure 4-2. The 3-Tier Universe

author's point of view or understanding of the natural world, we must always respect his or her original intention for writing a passage. Otherwise, we could make a passage mean whatever we wish for our own purposes.

To be sure, reading ancient texts like the Bible can be challenging and even surprising, as we have seen with verses like Genesis 1:2 and Philippians 2:10. The older a book is, the more difficult it will be for us to understand. This is because there is a greater conceptual distance between the intellectual context of ancient texts and that of modern readers.

These conceptual contexts are often called "hermeneutical horizons." The challenge for us as twenty-first century readers, being steeped in twenty-first century science, is to suspend our modern scientific ideas, and not to eisegetically force them into the Word of God. Therefore,

we need to read Scripture through ancient eyes

and with an ancient mindset.

THE BIBLE & ANCIENT SCIENCE

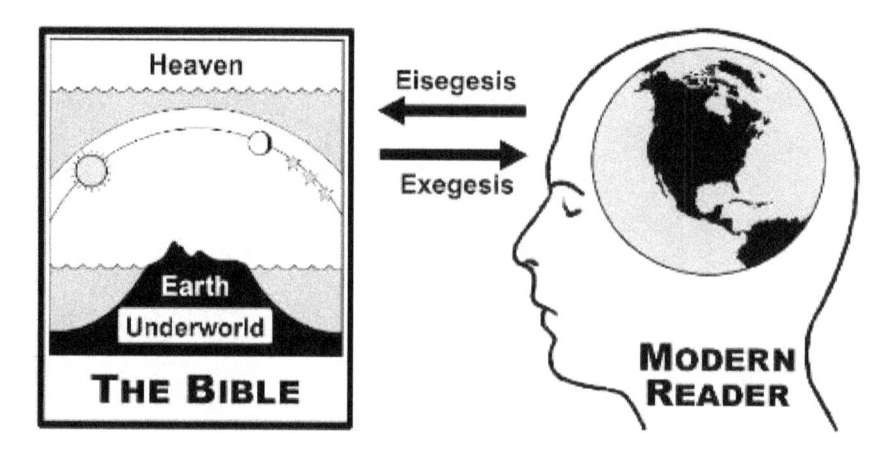

Figure 4-3. Hermeneutical Horizons & the Structure of the World

Figure 4-3 presents the hermeneutical horizons of the Bible and the modern reader with regard to the structure of the earth. This diagram also distinguishes between eisegesis and exegesis. For ancient people like the biblical writers, the universe was made up of three tiers with a flat earth. But for us today, we know the earth is spherical. Therefore, when we read the word "earth" in Scripture, there is a natural tendency for us to picture a sphere or globe. But that is eisegesis. Instead, we need to *recognize* and *respect* the ancient science in the Bible, even though we disagree with it. And we must practice exegesis and draw out from the Word of God the inspired writer's intended meaning.

Biblical Creation Accounts

To further illustrate the hermeneutical concepts of eisegesis and exegesis, let's look at how Martin Luther interpreted the structure of the heavens and the earth in Genesis 1. The cover of this book has a diagram of the universe found in his 1534 German translation of the Bible. It appears across from this first chapter of Scripture and the account of God creating the world in six days.

During Luther's generation the science-of-the-day was geocentrism. The Greek noun *gē* means "earth." This theory claimed that the earth is spherical and positioned at the center of the entire universe. It also asserted that the earth does not move. A sphere, termed the "firmament," enclosed the world and separated God and the heavenly realm from the rest of creation. Luther believed that the sun, moon, and stars were attached to the firmament, and the daily rotation of this heavenly sphere caused the sun to move around the earth, creating day and night.

Luther's sixteenth-century astronomy also appears in his 1536 biblical commentary *Lectures on Genesis*. With regard to the origin of heavenly bodies on the fourth day of creation, he writes, "Indeed, it is more likely that the bodies of the stars, like that of the sun, are round, and that they are fastened to the firmament like globes of fire."[2] In defending geocentricism and the immovability of the earth, Luther appeals to Joshua 10:12-13 and the miraculous stopping of the sun. This passage records, "Joshua said to the Lord in the presence of Israel: 'Sun, stand still over Gibeon' . . . The sun stopped in the middle of the sky and delayed going down about a full day." Luther argues, "I believe the Holy Scriptures, for Joshua commanded the sun to stand still, and not the earth."[3] In other words, Luther assumed the sun literally moved around the earth, and that it was the sun that was miraculously stopped by God in Joshua 10.

Now I am sure that you have identified two hermeneutical mistakes with Luther's interpretation of Scripture. First, the illustration of the universe in his 1534 translation of the Bible is eisegetical. Like most of my students who draw a sphere when picturing the earth in Genesis 1:2 (Fig. 4-1, p. 33), Luther forces his geocentric view of the world into Scripture. Second, Luther is a scientific concordist. He uses the Bible like a book of science. In attempting to argue that the sun actually moves across the sky, Luther reads Joshua 10:12-13 as a literal scientific statement to support the motion of the sun.

There are valuable lessons to be learned from Luther's hermeneutical mistakes (as well as our own!). I doubt there are many Christians today who believe in his geocentric understanding of the structure of the universe. And most of us do not think that the sun literally moves around the earth each day. Martin Luther demonstrates the problem with scientific concordism—Scripture cannot be aligned with science.

Another problem with concordism is that science changes over time. If one generation eisegetically forces their science into the Bible, then a later generation might discover these earlier scientific views are incorrect. And this is exactly what happened with Luther's geocentric interpretation of Scripture. No one today believes that the earth is at the center of the universe or that the sun is attached to a spherical firmament that rotates, moving the sun around the earth every day.

But there is a more serious problem with scientific concordism. Take for example the Christians who read Genesis 1 in Luther's Bible and saw the diagram of a geocentric universe across from this chapter. When it was later discovered that the earth moved around the sun, did these Christians lose their trust in Scripture? Or worse, did they lose their faith in the God of the Bible? Martin Luther's interpretive mistakes should serve as a warning to all of us that the Word of God should not be used as a book of science. Instead, the Bible reveals life-changing spiritual truths for developing a personal relationship with the God who inspired Holy Scripture.

HERMENEUTICAL PRINCIPLE 5

Ancient & Modern Phenomenological Perspectives

One of the most important lessons that I learned during my education in theology was to read the Bible through ancient eyes and with an ancient mindset. Yes, Scripture was written *for* every man and woman in every generation, but we also need to remember that it was written *to* a specific ancient people during a specific ancient time.[1] In particular, when interpreting statements about the natural world in the Bible, we must try to suspend our modern scientific concepts and think like an ancient person through their ancient understanding of nature. Let's look at some biblical verses to explain what I mean.

Scripture mentions that the sun moves across the sky. For example, Ecclesiastes 1:5 states, "The sun rises and the sun sets, and hurries back to where it rises." Psalm 19:4-6 asserts, "In the heavens God has pitched a tent for the sun. It is like a bridegroom coming out of his chamber, like a champion rejoicing to run his course. It rises at one end of the heavens and makes its circuit to the other." Similarly, Psalm 113:3 claims, "From the rising of the sun to the place where it sets, the name of the Lord is to be praised." And in Matthew 5:45, Jesus taught his disciples that God "causes his sun to rise on the evil and the good."

Many Christians are quick to explain the meaning of these verses referring to the movement of the sun. They argue that statements about nature in the Bible are from the perspective of what they look like to our natural senses, like the naked eye. The so-called "moving" sun in Scripture is only an appearance and merely a visual effect. More specifically, they contend that the biblical authors of these passages used phenomenological language. The Greek noun *phainōmenon* means "appearance." These Christians often add, "Check any weather app today and you will find the times for 'sunrise' and 'sunset.' Of course, everyone knows that these are just figurative terms and that the sun does not actually rise or set. Therefore, biblical verses referring to the sun moving across the sky should not be read literally." This is often referred to as the "phenomenological language argument."

On the surface, this seems to be a reasonable argument. But at a deeper level, we need to ask a key question: Did the God-inspired ancient writers of the Bible use phenomenological language in the same way that we use it today? In other words, when they referred to "sunrise" and "sunset," did they employ these terms in a figurative and nonliteral way like we do?

The answer to this question is "no." Proof comes from the history of science, and the conflict between the famous astronomer Galileo and the Catholic Church. Right up until the seventeenth century, nearly everyone including the scientific community believed that the sun literally moved across the sky daily and that the earth was immovable and stationary. But Galileo's work in astronomy with the newly invented telescope in 1609 opened the way to the idea that the earth rotated on its axis and revolved around the sun.[2] As a result, science discovered that the so-called "rising" or "setting" of the sun is only a visual effect caused by the rotation of the earth, giving the appearance (or phenomenon) that the sun "moves" across the sky every day.

The implications of this historical episode for biblical hermeneutics are significant. The divinely inspired writers of Scripture lived roughly 1500 to 3000 years *before* Galileo and his scientific discoveries. Therefore, these authors would have believed in the actual and literal movement of the sun and the immovability of the earth. So when Christians today argue that "sunrise" and "sunset" in Scripture are merely figurative terms and phenomenological language, they are eisegetically forcing their modern science into the Bible.

Now it is important to acknowledge that Scripture does employ phenomenological language in describing the natural world. However,

ANCIENT
Unaided
Physical Senses
Literal & Actual
Sun Rising/Setting

MODERN
Aided by
Scientific Instruments
Appearance of
Sun "Rising/Setting"

PHENOMENOLOGICAL PERSPECTIVES

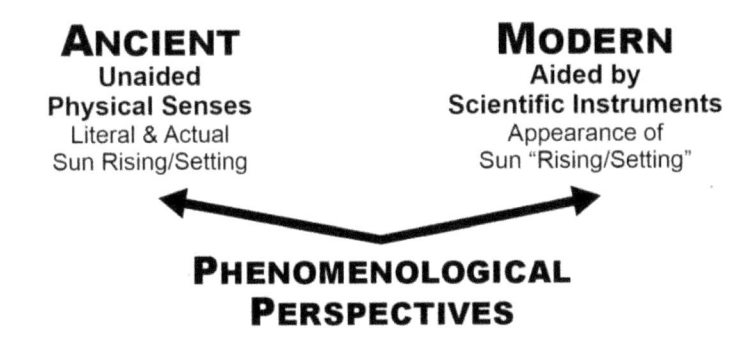

Figure 5-1. Ancient & Modern Phenomenological Perspectives

there is a critical and subtle difference between what the biblical writers saw and believed to be real in nature, and what we see and claim as a scientific fact. Observation in the ancient world was limited to unaided physical senses. What the authors of Scripture saw with their naked eyes, they believed to be real, like the literal rising and setting of the sun. Therefore, it is necessary to understand that statements in Scripture about nature are from an *ancient* phenomenological perspective.

In contrast, we view the world from a *modern* phenomenological perspective. We have the advantage of scientific instruments like telescopes, and these have widened and deepened our view of the physical world. When we see the sun "rising" and "setting," we know that this is only an appearance or visual effect caused by the rotation of the earth.

It is critical that we do not confuse and conflate (blend together) the ancient and modern phenomenological perspectives. This is a common mistake made by many Christians today. They assume that ancient people viewed and understood the natural world as we do. In particular, they presuppose that "sunrise" and "sunset" are non-literal figurative expressions in Scripture. However, this interpretation is a mistake. It is only in the seventeenth century that these became figures of speech. Therefore, the so-called "phenomenological language argument" is eisegesis. Figure 5-1 corrects this hermeneutical error by distinguishing between the ancient and modern phenomenological perspectives of the physical world.

But more importantly, I am certain that most Christians will agree that passages in Scripture referring to the movement of the sun are not intended to reveal scientific facts. Instead, this ancient astronomy is a vessel that delivers messages of faith.

For example, Psalm 113:3 states, "From the rising of the sun to the place where it sets, the name of the Lord is to be praised." Understood from an ancient phenomenological perspective, the places where the sun "rises" and "sets" are at the edges of a world with a flat earth. Therefore, this verse is calling all men and women everywhere to praise God. Similarly, in Matthew 5:45 Jesus teaches that the Creator "causes his sun to rise on the evil and the good, and sends rain on the righteous and the unrighteous." In other words, the Lord's kindness and sustenance extends to everyone on earth.

Biblical Creation Accounts

There is a question that I suspect has arisen in your mind. If the Bible has an ancient phenomenological perspective of astronomy, like the daily movement of the sun across the sky, then does it also have an ancient phenomenological perspective of biology? To answer this question let's try looking at living organisms through ancient eyes and think about them with an ancient mindset.

One of the best examples of ancient biology in the Bible is the categorization of bats as birds. Leviticus 11:13-19 records, "These are the birds you are to regard as unclean and not eat because they are unclean: the eagle, the vulture . . . [17 other birds are listed] . . . and the bat." Of course, this ancient understanding of taxonomy is reasonable because bats fly. However, they are mammals. Bats differ from birds in that they have body hair and mammary glands, and they develop in their mother's womb and not in an egg with a hard shell.

Another ancient taxonomical classification in Scripture is the idea that rabbits are ruminants that chew the cud. Leviticus 11:6 asserts, "The rabbit, though it chews the cud, does not have a divided hoof; it is unclean for you." But rabbits are not ruminants with multiple-chambered stomachs. Undoubtedly, the repetitive and side-to-side jaw actions of this animal gave the impression that they chew the cud like cows. Considering their limited understanding of anatomy and physiology, the taxonomical categories held by ancient people were quite logical. I am sure we would have accepted them had we lived at that time.

The Word of God also includes the ancient biological notion that living organisms are immutable. That is, they never change and remain the same. When ancient people looked at plants and animals, what would they have observed? They would have seen that wheat produced seeds that when sown only gave rise to wheat plants. Similarly, the seeds found inside fruit grew into trees that always produced the same fruit. People in the ancient world would also have noticed that hens lay eggs that only hatch chicks, female sheep continually give birth to lambs, and women are always the mothers of human infants. From an ancient phenomenological perspective, plants and animals were immutable and always remained constant.

In the light of this ancient perception of biology, it is now evident why the inspired ancient writer of Genesis 1 referred ten times to God creating plants and animals "according to their/its kinds." This was his observation and experience, and also that of everyone around him. Ancient people never saw a living organism change into an entirely different organism. And they were not aware of the evolutionary pattern in the fossil record. As a consequence, there is no hint of biological evolution in the Genesis 1 account of creation. Instead, plants and animals were seen as immutable, and it made perfect sense for ancient individuals to think that God had created each different kind of living creature separately and into its unique form.

This view of origins is called "*de novo* creation." The Latin preposition *dē* means "from" and *novus* is the adjective "new." *De novo* creation refers to the quick and complete origin of each living organism. This term is also used for the sudden creation of inanimate objects into fully formed structures like the earth, sun, moon, and stars. *De novo* creation appears in most ancient accounts of origins and features a Creator who

THE BIBLE & ANCIENT SCIENCE

acts dramatically through miraculous interventions.[3] This view of creation was the origins science-of-the-day in the ancient world, and it was clearly accepted by the Holy Spirit-inspired authors of Genesis 1 and 2. To be sure, there are some significant implications for the modern origins debate with regard to *de novo* creation in Scripture, and we will explore these as we proceed through our hermeneutical principles.

HERMENEUTICAL PRINCIPLE 6

The Message-Incident Principle

Let me now introduce the most important interpretive principle in this book on hermeneutics—the Message-Incident Principle as shown in Figure 6-1. It will help us understand passages in the Bible that refer to the physical world. I want to emphasize that this hermeneutical principle has a limited application. It is restricted to statements in Scripture that deal with nature, and it is in no way a concept that can be applied to every passage in the Bible. For example, this interpretive concept cannot be used with biblical texts dealing with the attributes of God such as his holiness (Rev. 4:8), Jesus' two great commandments (Matt. 22:37-40), or practices within the church like communion (1 Cor. 11:23-26).

I am convinced that most Christians already accept the basic idea behind the Message-Incident Principle in some implicit way. For instance, we all believe that the main purpose of the Bible is to reveal messages of faith and life-changing spiritual truths. I doubt that there are many Chris-

Bible
Statements about Nature

→ **MESSAGE**
Spiritual Truths
INERRANT

→ **INCIDENT**
Ancient Science
Ancient Phenomenological Perspective

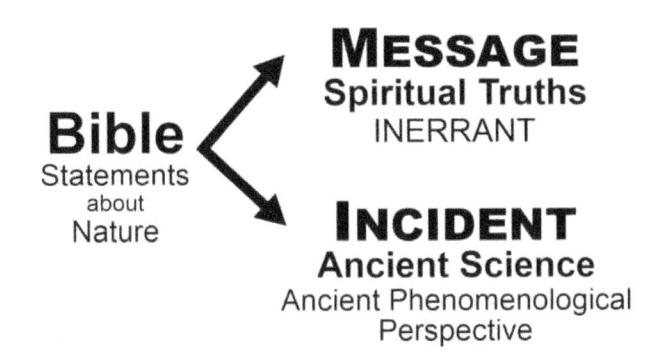

Figure 6-1. The Message-Incident Principle

tians who go to Scripture primarily to discover scientific facts about the natural world. Does anyone use Ecclesiastes 1:5, Psalm 19:4-6, or the words of Jesus in Matthew 5:45 as evidence that the sun literally moves across the sky and that every day it actually rises and sets?

First and foremost, the Message-Incident Principle asserts that spiritual truths in the Bible are *inerrant* because they are totally and absolutely true. The word "inerrant" means "to be completely free from error." Throughout history these messages of faith have consistently impacted the lives of men and women. They have assisted us in developing our personal relationship with the Lord and have provided joy, comfort, and purpose. The inerrant truths in Scripture are the foundational beliefs of the Christian faith. To use Genesis 1, the central messages of faith include: God is the Creator of the universe and life (v. 1), only men and women have been created in the Image of God (v. 26-27), and the marvellous world that God has made is very good (v. 31).

This fundamental hermeneutical principle also recognizes that statements in Scripture regarding the physical world feature an *ancient science*. More specifically, the inspired biblical writers and their readers understood nature from an ancient phenomenological perspective. They did not enjoy sophisticated scientific instruments like telescopes and microscopes as we do today. Their view of the creation was limited to their natural senses, such as observation through the naked eye. Nevertheless, Scripture features the best science-of-the-day in the ancient world of the biblical peoples. Had we lived at that time, we would have embraced their ancient scientific ideas, like the literal movement of the sun across the sky every day.

The Message-Incident Principle states that the ancient science in Scripture is *incidental* because God's central purpose in the Bible is to reveal messages of faith, and not scientific facts about his creation. The word "incidental" has the meaning of "that which happens to be alongside" and "happening in connection with something more important." In this way, the ancient science in Scripture is found "alongside" the "more important" inerrant spiritual truths revealed by the Holy Spirit to the biblical writers.

Though the ancient science in the Bible is ultimately incidental to the messages of faith, it plays a *critical role* in delivering these spiritual truths. The ancient scientific ideas are similar to a cup that holds water. Does it really matter whether a cup is made of glass, plastic, or metal? No. The material that it is made of is incidental. What matters is that a vessel is needed to bring water to a thirsty person. Similarly, the incidental ancient science in Scripture is like a cup that delivers the life-giving spiritual messages to our thirsty souls.

Let's apply the Message-Incident Principle to Philippians 2:9-11. As we noted, a more precise translation of these verses by the apostle Paul states, "Therefore God exalted him [Jesus] to the highest place and gave him the name that is above every name, that at the name of Jesus every knee should bow, [1] in heaven and [2] on earth and [3] down in the underworld, and every tongue confess that Jesus Christ is Lord, to the glory of God the Father." The inerrant spiritual truth in Philippians 2:9-11 is clear: Jesus is the Lord over the entire creation. In order to reveal this message of faith to Paul and his ancient readers, God allowed the incidental ancient science of the 3-tier universe to be used as a vessel to deliver it.

Now I suspect there are some of you who are probably asking the question, "Did God lie in the Bible?" My answer to this question is an emphatic "NO!" In fact, Scripture states quite clearly in Titus 1:2 that God "does not lie," and Hebrews 6:18 asserts that "it is impossible for God to lie." Lying requires an individual to be deceptive, and the God of the Bible is certainly *not* a God of deception.

The God of Christianity is a God of truth and love. To reveal himself to an ancient people, he graciously came down to their intellectual level to communicate his life-changing spiritual truths. In Philippians 2:9-11, the Holy Spirit allowed the apostle Paul to use the ancient notion of the 3-tier universe as an incidental vessel to deliver an inerrant mes-

THE BIBLE & ANCIENT SCIENCE

sage of faith. As a result, Paul and his readers would have fully comprehended that Jesus is the Lord of the whole world, because from their ancient phenomenological perspective, the 3-tier universe was understood to be the entire universe.

Of course, God could have revealed to Paul modern scientific concepts like spiral galaxies, solar systems, and neutron stars, etc. But do you think that this apostle and his readers would have understood what these astronomical structures were? I doubt it. They did not have powerful telescopes as we do today. Such a revelation would have been confusing to ancient people and most likely a stumbling block that would have stopped them from embracing the inerrant spiritual truth of Jesus' lordship over the entire creation. Therefore, God did *not* lie in Philippians 2:9-11. Instead the Holy Spirit graciously accommodated and descended to the level of ancient men and women in the process of inspiring the Bible.

The Message-Incident Principle also assists us to separate the inerrant spiritual truth in Philippians 2:9-11 from its ancient science, and then to recast this message for our twenty-first century generation by using modern science as an incidental vessel. For example, as Christians today we can proclaim that Jesus is the Lord of our massive 13.8-billion-year-old universe with its approximately 100 billion galaxies featuring about 100 billion stars in each galaxy! As science advances, every amazing discovery in nature can be viewed in the light of God's lordship over his creation.

Finally, the Message-Incident Principle sheds light on a problem that appears regularly within our churches. Most Christians are not aware that the Bible has an ancient understanding of science. They assume that statements about nature in Scripture align with physical reality. By embracing scientific concordism, they often conflate the inerrant spiritual truths in the Bible with the incidental ancient science. The term "conflate" refers to "the careless blending or mixing of distinct ideas." In this way, many Christians believe that statements about nature in Scripture are inerrant truths. To correct this situation, the Message-Incident Principle helps us to *separate* the inerrant messages of faith from the incidental ancient science, and to not *conflate* the two together.

Biblical Creation Accounts

In a manner similar to Philippians 2:9-11 presented above, we can apply the Message-Incident Principle to the creation accounts in Scripture. As we noted in the previous hermeneutical principle, the biblical authors and their readers were very logical in believing that plants and animals were immutable. It was also quite reasonable for them to think that God had created living organisms *de novo* (quick and complete) "according to their/its kinds," as stated ten times in Genesis 1.

Therefore, the message of faith in this biblical creation account is that the God of the Bible is the Creator of every plant and every animal. In order to deliver this inerrant spiritual truth, the Holy Spirit descended to the level of the biblical writers and allowed their incidental ancient science of *de novo* creation to be used as a vehicle to transport this foundational belief to the ancient readers of this first chapter in Scripture.

Of course, many Christians today believe that God's *de novo* creative acts in Genesis 1 are a record of actual historical events in the origin of living organisms. Young earth creationists contend that plants were created rapidly and fully formed on the third day of creation, birds and sea creatures on the fifth day, and land animals and humans on the sixth. And each of these days were 24-hour periods. Progressive creationists also believe that the Creator made living creatures quickly and completely. They claim that these miraculous *de novo* creative events occurred at different times during the 4.6-billion-year history of the earth. According to this view of origins, the days of Genesis 1 are periods that are millions of years long (See Appendix 1).

I am certain that you have identified the problem with these two Christian anti-evolutionary views of origins. Young earth creation and progressive creation are scientific concordist positions that conflate the ancient science of *de novo* creation with the inerrant message of faith that God created all living organisms. This would be no different than to take

THE BIBLE & ANCIENT SCIENCE

the spiritual truth in Philippians 2:9-10—Jesus is the Lord of the entire universe—and to conflate it with the ancient science of the 3-tier universe, and then to claim that we must accept this ancient understanding of the structure of the world. I doubt that any Christian today would embrace such a position.

The Message-Incident Principle underlines that we must not conflate the inerrant spiritual truths in Scripture with the incidental ancient science that transports them. Instead, we need to separate the two in order to focus on God's intended messages of faith for us. It is worth noting that biblical interpreters throughout history have often conflated ancient concepts of nature in Scripture with God's messages of faith without being aware of it. The reason for this is that the identification of ancient science in Scripture can only occur *after* the discovery of modern scientific concepts.

For example, it was only after Galileo's work in astronomy during the seventeenth century that Christians realized biblical passages referring to the sun's movement across the sky were based on an ancient phenomenological perspective. This historical episode is one of the reasons why we as Christians must keep up to date with the latest scientific discoveries—it allows us to be better interpreters of the Word of God.

Excursus
Are the Messages of Faith Merely Ancient Human Beliefs?

When I introduce the Message-Incident Principle to my science and religion students, they are quick to challenge me with several questions. If the science in the Bible is an ancient human understanding of nature, then is this also the case with the spiritual truths in Scripture? Since no one today accepts ancient science like the 3-tier universe in Philippians 2:9-11, why should we believe the message of faith in this passage that Jesus is Lord of the entire cosmos? And are we not being inconsistent if we reject the ancient phenomenological perspective of the world in the Bible but accept the spiritual truths? I suspect that many of you are probably asking the same important questions.

My response to my students is simple and rather obvious. I say to them: The fact that you are sitting here in a class on Christian theology two thousand years after the Bible was composed is proof of the power and eternal truthfulness of the messages of faith. The ancient science in Scripture such as the 3-tier universe is not the reason you are in my course. For that matter, before you entered my classroom, most of you were not aware that the Word of God includes this ancient understanding of the structure of the world.

Instead, it is inerrant messages of faith, such as the divine revelation in Philippians 2:9-11 that Jesus is the Lord of the whole universe, that have led you to become a Christian. In fact, it is the power of the spiritual truths in Scripture that has caused men and women throughout history to be born-again and to change their lives in dramatic ways. This reality of the impact of Scripture on humans is clearly stated in Hebrews 4:12. "For the Word of God is alive and active. Sharper than any double-edged sword, it penetrates even to dividing soul and spirit, joints and marrow; it judges the thoughts and attitudes of the heart." And this indeed is my personal experience when reading the Bible.

If the messages of faith in Scripture were merely ancient human ideas about spirituality similar to those in other ancient religions, then they should have died away like most of these religions a long time ago. Take for example some of the religious beliefs in ancient Mesopotamia.[1] In one creation account, a god murders a goddess and then splits her body in half to make heaven and earth. The reason humans are created in many of these stories is to relieve the gods of their work. One justification for a worldwide flood in some Mesopotamian accounts is that humans were too noisy and the gods could not sleep. During the flood the gods suffer from hunger because there are no humans to feed them sacrifices. I think everyone will agree that the gods in these Mesopotamian stories are quite pathetic and the spiritual truths just as sorrowful. It is not surprising that these religious beliefs passed away and have had no influence on later generations.

In sharp contrast, the God of the Bible is a majestic, powerful, and holy God. He is in complete control of the universe with no other gods in existence to challenge him. The Lord does not require men and women to meet his needs. As Acts 17:25 states, God "is not served by human hands, as if he needed anything." The Lord values humans and creates us in his likeness and image (Gen. 1:26-27). Amazingly, the Creator of the entire universe is in a personal relationship with us. And human sin is the reason for divine judgment. The attributes of the God of Scripture—such as holiness, love, and truthfulness (Rev. 4:8; 1 Jn. 4:8; Heb. 6:18)—are so far above and beyond the attributes of the pagan gods of nations that surrounded ancient Israel and the earlier Christians. In fact, there is no comparison between our God and their gods.

Moreover, what is quite remarkable about the Bible is that God began to reveal himself to humanity through a small and insignificant tribal nation like Israel, and not a major civilization such as the Mesopotamians. Jesus then used twelve mostly uneducated men as disciples to preach the gospel that he died for the sins of men and women. The Lord did not employ the powerful Romans or Greeks. If the biblical messages of faith were merely ancient human ideas of irrelevant and inconsequential ancient people, then they should have disappeared along with these small communities and never gained prominence around the world. However, the spiritual truths in the Bible are "alive and active" (Heb. 4:12), and they have deeply struck human souls throughout history and continue to do so today.

Proof that the messages of faith in Scripture are not just ancient human beliefs about religion is demonstrated by the fact that you are reading a book on biblical hermeneutics. For me, this means that the Bible has impacted you in a very profound way and that you want to improve your interpretation of the Word of God. It is not the ancient scientific idea of a 3-tier universe that has led you to a personal relationship with Jesus. Rather, it is the eternal and inerrant spiritual truth that Jesus is Lord of the entire world that has powerfully changed your life. Do you agree?

HERMENEUTICAL PRINCIPLE 7

Biblical Accommodation

The verb "to accommodate" has the meanings "to adapt," "adjust," "help out," and "make fit." Within the context of biblical hermeneutics, the principle of accommodation refers to God adapting his revelation to the level of humans in order that we may understand his inerrant spiritual truths. To explain this principle, let's look at three passages where Jesus uses an ancient view of botany as an incidental vessel to deliver messages of faith.

In the well-known parable of the mustard seed, the Lord asks his disciples in Mark 4:30-32, "What shall we say the kingdom of God is like, or what parable shall we use to describe it? It is like a mustard seed, which is the smallest of all seeds on earth. Yet when planted, it grows and becomes the largest of all garden plants, with such big branches that the birds can perch in its shade."

Is the mustard seed "the smallest of all the seeds on the earth"? No. Orchid seeds are much smaller. So did Jesus make a mistake? Or worse, did he lie to his disciples? No, not at all! Instead the Lord was adapting or accommodating his message about the kingdom of God to his ancient audience. In other words, he was using the botany-of-the-day, which for his listeners included the idea that the mustard seed was the tiniest of all plant seeds on the earth. The parable in Mark 4:30-32 is prophetic. The kingdom of God would begin with a small number of disciples, and then grow into a great body of believers—the church.

Jesus offers another teaching about God's kingdom in the parable of the growing seed. In Mark 4:26-29 he states, "This is what the kingdom of God is like. A man scatters seed on the ground. Night and day, whether he sleeps or gets up, the seed sprouts and grows, though he does not know how. All by itself the soil produces grain—first the stalk, then the head, then the full kernel in the head. As soon as the grain is ripe, he puts the sickle to it, because the harvest has come."

As everyone knows today, the soil does not produce grain "all by itself" because DNA in the seed is a significant contributor in the growth of a plant. Once again, Jesus did not lie to his disciples, nor did the Holy Spirit make a mistake while inspiring the biblical writer of the Gospel of Mark. This is another example of God accommodating to the level of ancient people by using their understanding of botany to deliver an inerrant spiritual truth. Mark 4:26-29 reveals that though we do not fully understand the spiritual growth of those in the kingdom of God, the Lord will gather us together at the end of time.

In John 12:23-24, Jesus speaks to disciples about his imminent death and resurrection. "The hour has come for the Son of Man to be glorified. Very truly I tell you, unless a kernel of wheat falls to the ground and dies, it remains only a single seed. But if it dies, it produces many seeds." Modern science has discovered that seeds are alive and functioning at a very low metabolic rate. If a seed died, it would never germinate.

Think about what happens to the outer shell of a seed prior to germination. It breaks down and rots. Therefore, from an ancient phenomenological perspective, it was reasonable for ancient people to assume that seeds die before they germinate. In John 12, Jesus used this ancient botanical idea as a vessel to deliver the message of faith that he would be put to death, and that later he would be resurrected physically from the grave, leading many people to have faith in him.

I need to make a few more comments about these passages with Jesus employing an ancient knowledge of plant seeds. Most importantly, I must emphasize that the Lord did not lie. As we noted in the previous interpretive principle, Titus 1:2 asserts that God "does not lie" and Hebrews 6:18 states that "it is impossible for God to lie." Jesus had no intention whatsoever to deceive anyone. Instead, in order to reveal spiritual truths at the level of his ancient listeners, he accommodated by using the botany-of-the-day in his teaching.

We must also remember that Jesus is God and that he is the Creator of the world. John 1:3 records that through Jesus "all things were made; without him nothing was made that has been made." Similarly, in referring to the Lord, Colossians 1:16 states that "all things have been created through him and for him." And Hebrews 1:2 records that God spoke through Jesus, "whom he appointed heir of all things, and through whom also he made the universe." Being the Creator of everything including plants, Jesus certainly knew that the mustard seed is not the smallest seed, the soil is not the only factor in plant growth, and seeds do not die before germination.

Finally, I think that most Christians would agree that the Lord did not come to earth to teach scientific facts about plants and their seeds! Instead, in Mark 4:26-29, Mark 4:30-32, and John 12:23-24, Jesus used the botany-of-the-day that was familiar to his listeners in order to communicate messages of faith as effectively as possible.

It must be mentioned that some Christians complain that the principle of biblical accommodation "weakens" or "waters down" the Scriptures. I certainly appreciate their concern. But here are six reasons for accepting the belief that accommodation is a feature of the Word of God.

First, divine revelation by necessity requires God to accommodate. Stated another way, for an Infinite Creator to communicate with finite creatures, he must come down to our level. Otherwise we would never understand. The mind of God is so much greater than the mind of humans. As the Lord states in Isaiah 55:8-9, "For my thoughts are not your thoughts . . . As the heavens are higher than the earth . . . so are my thoughts higher than your thoughts."

Second, the greatest act of divine revelation is the Incarnation. The Latin noun *carnis* means "flesh." John 1:14 states that Jesus "became flesh and made his dwelling among us." Similarly, Philippians 2:7-8 asserts that the Lord "humbled himself" and "made himself nothing" in order to become a man. In other words, God accommodated by taking on human flesh, and through this ultimate act of accommodation, the Lord revealed his unfathomable love for us by dying on the Cross for our sins.

THE BIBLE & ANCIENT SCIENCE

Third, Jesus often employed parables in his teaching. Simply defined, parables are earthly stories with heavenly messages. The Lord included "earthly" ancient scientific notions-of-the-day, like the mustard seed being the smallest of all seeds, in order to deliver inerrant spiritual truths about the kingdom of God.

Fourth, the Lord was fully aware of the limitations of humans. Following his parables of the mustard seed and growing seed, Mark 4:33 comments, "With many similar parables Jesus spoke the word to them, *as much as they could understand*" (my italics). Clearly, the Lord accommodated to the intellectual level of his disciples and audience in order to teach his messages of faith.

Fifth, Christians experience accommodation when they pray. Most would agree that God meets us exactly where we are at in our life, and that he talks to us at our level of comprehension. As we grow spiritually over time, the Lord then reveals himself using spiritual beliefs that are more mature. God knows us better than we know ourselves and knows the best way to communicate his inerrant truths so that we can fully grasp them.

Finally, everyone at some point uses accommodation because it is an effective and natural way to communicate. For example, when a 4-year old asks where babies come from, parents answer by coming down both physically to their knees and intellectually to the level of understanding of the child. They communicate the message of faith that a baby is a gift from God without presenting the incidental details of sexual reproduction. In other words, *it is possible to reveal inerrant spiritual truths without using actual scientific facts.*

Biblical Creation Accounts

Here are a few questions I would like you to think about regarding biblical accommodation. Are the statements about origins in Genesis 1 and 2 similar to the incidental ancient botany used by Jesus in his teaching? Did the Holy Spirit accommodate by allowing these biblical writers to use their ancient view of origins? And is it possible that God revealed inerrant spiritual truths in Genesis 1 and 2 without employing modern scientific facts about how he actually created the universe and living organisms?

In my opinion, the principle of accommodation is one of the most important hermeneutical concepts for interpreting statements in Scripture that deal with the physical world. The Bible is the Holy Spirit-inspired Word of God. As 2 Timothy 3:16 states, "All Scripture is God-breathed and useful for teaching, rebuking, correcting and training in righteousness." But we must always remember that the inspired biblical writers were ancient people. In order to reveal the inerrant spiritual truth that God was the only Creator of the world, the Holy Spirit had to descend to their level of comprehension and accommodate by using the origins science-of-the-day.

Let me further explain. Assume for a moment that you are God and that you have decided to reveal yourself to the world through an ancient community like the Hebrews about 3500 years ago. Being God, you have the power to inspire writers to record your every word and thought in Scripture. And let's also assume that you created living organisms through an evolutionary process. In your creation account, would you have inspired the biblical author to write: "In the beginning God created plants and animals through evolution."

I doubt that ancient men and women thousands of years ago would have understood the meaning of the term "evolution." The theory of biological evolution was only accepted after the scientific discoveries of Charles Darwin in the late nineteenth century. In fact, many surveys of American adults today show that even though we are the most scientifically informed generation that has ever lived, about half of us reject evolution.[1] So it seems highly unlikely that ancient people would have understood a creation account about God creating living organisms through an evolutionary process.

Instead, I suspect that if you were God and wanted to reveal to an ancient community that you were the Creator of plants and animals, you would use their understanding of origins. From an ancient phenomenological perspective, they saw that living organisms were immutable and never changed. It was perfectly logical for them to think that these creatures were created *de novo* (quick and complete). Therefore, by accommodating to the intellectual level of ancient people, would you not say that you created plants and animals rapidly and fully formed? And isn't this exactly what the Holy Spirit did by revealing that the God of Christianity was the Creator of life in Genesis 1 and 2?

HERMENEUTICAL PRINCIPLE 8

Authorial Intentionality: Divine & Human

When authors write articles or books, they intend to communicate some specific ideas to their readers. For example, my primary intention in this book is to offer a way to interpret passages in Scripture that refer to the natural world and the biblical creation accounts. However, the Bible is unique in that it is authored by both the Holy Spirit and human writers. As a result, Scripture features a divine authorial intention and a human authorial intention.

The Bible affirms its dual authorship and intentionality. Hebrews 1:1 states, "In the past God spoke to our ancestors through the prophets at many times and in various ways." In Numbers 12:6, God declares, "When there is a prophet among you, I, the Lord, reveal myself to them in visions, I speak to them in dreams." Similarly, Acts 3:21 claims that "God spoke by the mouth of his holy prophets from ancient time" (NASB). As well, 2 Peter 1:20-21 asserts that "prophets, though human, spoke from God as they were carried along by the Holy Spirit." And referring to King David's prophecy in Psalm 110:1, Jesus affirms in Mark 12:36 that "David himself speaking by the Holy Spirit declared . . ."

Let's revisit some biblical passages that deal with the natural world and view them in the light of the interpretive principle of authorial intentionality. As we noted previously, a more accurate translation of Philippians 2:9-11 states, "Therefore God exalted him to the highest place and gave him the name that is above every name, that at the name of Jesus every knee should bow, [1] in heaven and [2] on earth and [3] down in the underworld, and every tongue confess that Jesus Christ is Lord, to the glory of God the Father."

The apostle Paul's primary intention in this passage was to assert that Jesus is the Lord of the entire universe. From the ancient phenomenological perspective of this human author and his generation, the 3-tiered universe was the best understanding of the structure of the world at that time. The central intention of the Holy Spirit in Philippians 2:9-11 was also to reveal the lordship of Jesus over the whole creation. In order to do so, the Divine Author accommodated to the level of Paul and his readers by allowing an ancient astronomy and ancient geography to be used as an incidental vessel in delivering this inerrant spiritual truth.

Genesis 1:11 records the origin of land plants on the third day of creation, and Genesis 1:24 presents the creation of land animals on the fourth day. "Then God said, 'Let the land produce vegetation: seed-bearing plants and trees on the land that bear fruit with seed in it, according to their various kinds.'. . . And God said, "Let the land produce living creatures according to their kinds: the livestock, the creatures that move along the ground, and the wild animals, each according to its kind.'"

The central intention of the writer of Genesis 1 in these two verses is to state the inerrant message of faith that the God of the Bible created the plants and animals that live on dry land. In doing so, this human author employs *de novo* creation, which was the best origins science-of-the-day. Likewise, in Genesis 1:11 and 24, the Holy Spirit intended primarily to reveal the identical spiritual truth. Yet the Divine Author accommodated and allowed the use of the ancient scientific notion that God created land plants and animals quickly and completely mature.

It is evident from the examples of Philippians 2:9-11 and Genesis 1:11 and 24 that the divine authorial intention is the same as the human authorial intention. The spiritual truths of the Holy Spirit and the biblical writer are identical in each passage. This is also the case with the ancient science in each of these two passages. For the human author, this was the science that was accepted by his generation at that time; for the Divine Author, allowing ancient understandings of nature in the Bible was an act of accommodation. There is no evidence in Philippians 2:9-11 and Genesis 1:11 and 24 that God intended to reveal modern scientific ideas to the biblical writers.

Biblical Creation Accounts

The accounts of origins in the Book of Genesis feature both a divine and a human authorial intentionality. For example, Christians throughout the ages have agreed that Genesis 1 reveals the inerrant spiritual truths that God created the universe and life, the world is very good, and only men and women are created in the Image of God. The primary intention of both the Holy Spirit and the biblical author was to disclose these three main messages of faith.

Even though Christians embrace these eternal truths, they continue to disagree about whether or not God intended to reveal modern scientific facts in Scripture ahead of their discovery by modern science. In other words, there is debate in the church over divine authorial intentionality and scientific concordism. To explore this notion, let's examine the interpretation of Genesis 1:1 by two leading scientific concordists.

As we have noted earlier, Henry Morris is the father of modern young earth creation. In explaining Genesis 1:1, he claims:

> The universe, therefore, in essence must be a continuum of Space, Time and Matter-Energy. No one of the three can exist without the other. Therefore, the entire continuum must have existed simultaneously from the beginning. This fundamental truth is taught explicitly in the first verse of the Bible, which is the foundation of everything else. "In the beginning (Time), God created heaven (Space) and earth (Matter)" Genesis 1:1. The foundational verse of the Bible thus yields the foundational fact of science![1]

This passage by Morris raises two important questions regarding authorial intentionality. First, if the Holy Spirit's intention was to reveal the twentieth century scientific notion of the space-time-matter/energy continuum in Genesis 1:1, then what was the ancient human writer's intention in this verse? Undoubtedly, it was not this modern idea of science, because there is no evidence in the ancient world that anyone was aware of such an incredibly complex concept. But as we noted above with Philippians 2:9-11 and Genesis 1:11 and 24, Scripture features a dual authorial intentionality in which the Holy Spirit and the biblical writer employed the same ancient science, and not modern science.

Second, did the ancient human writer of Genesis 1:1 intend the word "heaven" to refer to "space" and the term "earth" to mean "matter," as suggested by Henry Morris? Again, this is doubtful. In the ancient world of the biblical authors, "heaven" referred to a region above the dome of the sky, and "earth" was understood to be flat with an underworld, as shown in Figure 4-2 on page 35. The notions of "space" and "matter" are concepts used by twentieth century physicists. It is obvious that Morris is eisegetically forcing modern scientific ideas into the Word of God.

Hugh Ross is the leading progressive creationist today. He offers the following concordist interpretation of Genesis 1:1.

> 'In the beginning God created the heavens and the earth' (Genesis 1:1) ... The Bible says in unequivocal terms that the "heavens and the earth" began, that they exist for finite time only, and that God exists and acts inside, outside, and before the universe's space-and-time boundaries ... New scientific support for a hot big-bang creation event, for the validity of the space-time theorem of general relativity, and for ten-dimensional string theory verifies the Bible's claim for a beginning. In the final decade of the twentieth century, astronomers and physicists have established that all of the matter and energy in the universe, and all of the space-time dimensions within which the matter and energy are distributed, had a beginning in finite time, just as the Bible declares.[2]

Here again we have an attempt to align the Bible with modern science. But Hugh Ross fails to recognize and respect that statements about nature in Scripture feature an ancient science. As noted above with Henry Morris, the "heavens and the earth" in Genesis 1:1 refers to a place above the dome of the sky and a flat earth in a 3-tier universe, and not to the "universe's space-and-time boundaries." Moreover, it is highly doubtful that the human author of this verse intended to deal with the "big-bang," "general relativity," and "ten-dimensional string theory." Once again, this forcing of twentieth century scientific concepts into the Word of God is eisegesis.

Let me end this hermeneutical principle with a rather blunt comment that I believe needs to be made. Why is it that Christianity had to wait until our generation and the discovery of complex scientific concepts like space-time-matter continuum before we could interpret Genesis 1:1 correctly, as suggested by Henry Morris and Hugh Ross? I am sorry to say, but this strikes me as rather self-centered and self-serving. Why should we think that we are the first and only Christians to come to the true interpretation of the first verse of the Bible? Is my criticism of these two modern day scientific concordists too harsh, or is it reasonable? You decide.

HERMENEUTICAL PRINCIPLE 9

Biblical Sufficiency & Human Proficiency

Every Christian knows that interpreting the Bible can certainly be difficult at times. Yet despite hermeneutical challenges, the power of the Word of God is demonstrated by the fact that those seeking the Lord are always able to meet him, and they are also capable to discern his inerrant messages of faith in Scripture. As Psalm 9:10 states, "For you, Lord, have never forsaken those who seek you." In this way, the Bible is a sufficient revelation of God that allows us to develop a personal relationship with him, and human readers are proficient in grasping its life-changing spiritual truths.

It is crucial to emphasize that there is a spiritual and personal component to biblical interpretation. Reading and understanding the Word of God is not just an intellectual exercise, and it is not limited to pastors and theologians only. Scripture is for everyone. But in order to fully grasp the meaning of a biblical passage, we must get down on our knees before our Creator and open our hearts in order to listen to him speak to us through his Word. In particular, we need to have a submissive attitude before the Lord. And it is only with this humble and obedient disposition that the spiritual truths in Scripture can lead to "refreshing the soul," "making wise the simple," and "giving joy to the heart" (Ps. 19:7-8).

Reading the Bible is a mystical experience. It is a spiritual encounter between us and the Lord, facilitated by the inspired words in Holy Scripture. As we noted previously in Hebrews 4:12, "For the Word of God is alive and active. Sharper than any double-edged sword, it penetrates even to dividing soul and spirit, joints and marrow; it judges thoughts and attitudes of the heart." And for those of you who have had this powerful experience of meeting God in his Word, you know that there is a "voice" that transcends the words on the pages of Scripture. It is the voice of the Holy Spirit that speaks to us in our heart and mind (Jer. 31:33; Heb. 10:16).

Church history affirms the hermeneutical principle of the sufficiency of the Bible and the proficiency of its readers. Take for example our understanding of Jesus in Scripture. The vast majority of Christians throughout history have been united by the belief that he is indeed God who became a human. Despite the many sorrowful conflicts between the three main branches of Christianity (Catholic, Orthodox, and Protestant), they steadfastly embrace the central inerrant message of faith in the New Testament—the Incarnation. As John 1:14 states regarding Jesus, "The Word became flesh and made his dwelling among us."

The power of this inerrant spiritual truth is also confirmed by history. Men and women who have believed that Jesus is God in the flesh have been dramatically born-again and led to change their life in radical ways. Personally, I have had this powerful spiritual experience. In sharp contrast, a small number of individuals at different times have believed that Jesus was just a man with some novel religious ideas. However, this liberal theology has never captured the souls of the great majority of people seeking to find God. It has titillated the intellectual curiosity of some intellectuals, but rarely does this view of Jesus lead to spiritual transformation, or has it inspired many to follow his call to "go and make disciples of all the nations, baptizing them in the name of the Father, and of the Son and of the Holy Spirit" (Matt. 28:19).

A subtle implication of the principle of biblical sufficiency and human proficiency is that there are aspects of Scripture that do not contribute to the development of a personal relationship with the Lord. An example would be the incidental ancient understanding of the natural world in the Bible. I doubt anyone comes to Christ because of the 3-tier universe in Philippians 2:10. Instead, it is the mystical experience of believing that Jesus is Lord of the entire world that leads men and wom-

THE BIBLE & ANCIENT SCIENCE

en to him. To further demonstrate this point, let's again consider Genesis 1:1, "In the beginning God created the heavens and the earth." Christians throughout time have envisioned this verse a wide variety of ways.

The oldest interpretation and depiction of the first verse of the Bible is that of a 3-tier universe, which we saw in Figure 4-2 on page 35. This understanding of the structure of the world has heaven above the dome of the sky, a flat earth in the middle, and an underworld below. As we have noted, Philippians 2:10 in the beloved Kenotic Hymn indicates that the apostle Paul believed the cosmos was made up of three tiers.[1]

For roughly the first three-quarters of church history, many important Christians accepted geocentrism. The Greek noun *gē* means "earth." According to this view, the earth is spherical and at the center of the entire universe. There were two versions of this theory. The first contended that heaven was a single sphere that enclosed the earth as presented in Figure 9-1. This heavenly sphere was termed the "firmament," and the sun, moon, and stars (fixed and wandering) were placed in it. As we saw, the famous sixteenth century theologian Martin Luther accepted this structure of the world as seen on the cover of this book.

Many Christians accepted a second geocentric theory. It pictured the earth surrounded by a series of spheres, with each sphere having a plan-

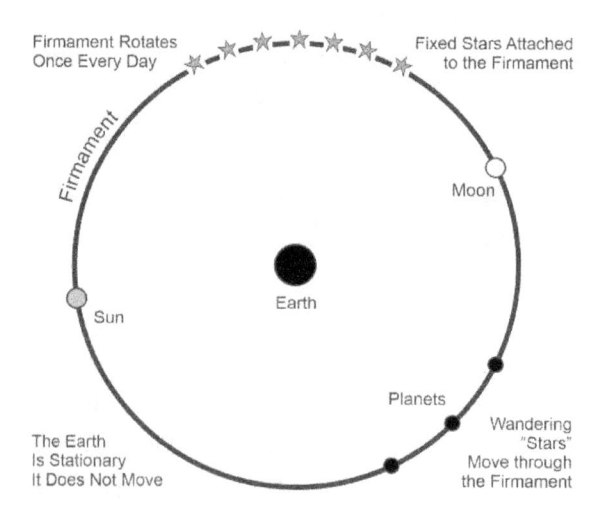

Figure 9-1. Geocentric Universe with a Single Heavenly Sphere

BIBLICAL SUFFICIENCY & HUMAN PROFICIENCY

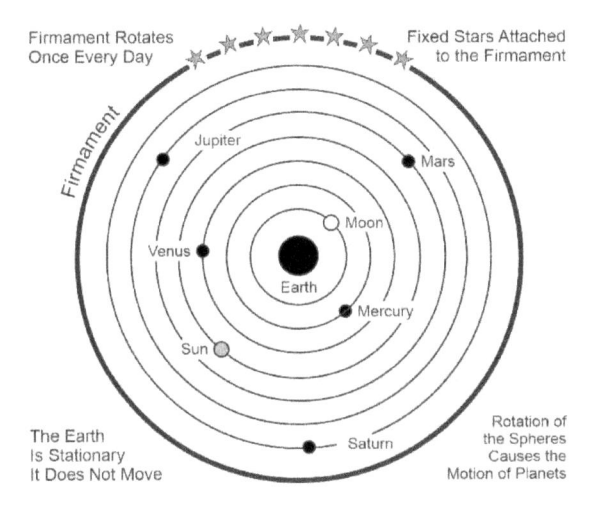

Figure 9-2. Geocentric Universe with Multiple Heavenly Spheres

et, as shown in Figure 9-2. The moon was also in a sphere. According to this view, the motion of the spheres caused the motion of the sun, moon, and planets.[2] The well-known sixteenth century theologian John Calvin held this view of the cosmos.

During the seventeenth century, heliocentrism eventually became popular within the church. The Greek noun *hēlios* means "sun." As Figure 9-3 illustrates, this theory places the sun in the middle of the entire universe. It

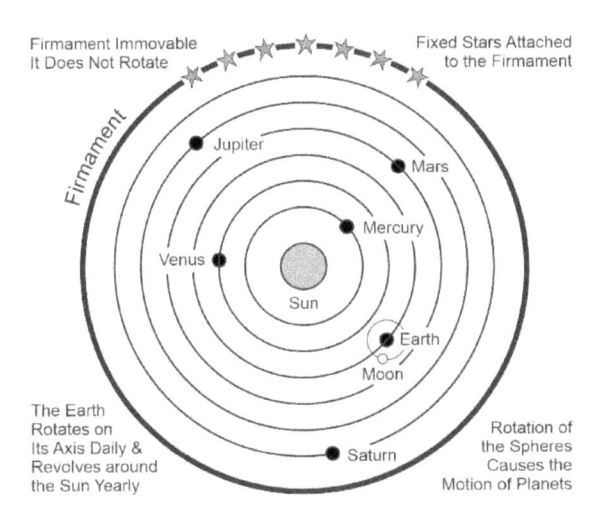

Figure 9-3. Heliocentric Universe

also contends that the movement of spheres in heaven produced the motion of the planets and the moon. This view of the world was proposed by the famous astronomer Nicolas Copernicus in the sixteenth century, and it was popularized by Galileo in the first half of the next century.

Finally, as twenty-first century Christians, we accept modern astronomy. The universe is about 5,500,000,000,000,000,000,000,000 miles wide, with around 100 billion galaxies and approximately 100 billion stars in each galaxy! Our earth rotates on its axis and revolves around the sun in a solar system that is part of the Milky Way Galaxy. And of course, we do not believe that astronomical bodies are attached to moving spheres. Instead, gravity accounts for the motion of galaxies, stars, planets, and moons.

Even though Christians throughout the ages have interpreted and pictured Genesis 1 in a wide variety of ways, they stand united in believing the central spiritual truth of this verse—God made the heavens and the earth. This is solid evidence that the Bible has been sufficient in revealing the world has a Creator, and Christians have been proficient in drawing out this foundational message of faith from Scripture, regardless of their scientific views.

Biblical Creation Accounts

Are the accounts of origins in Genesis 1-3 a sufficient divine revelation for us to meet our Creator if we earnestly seek him? And with an open heart, are we proficient in reading and understanding his main spiritual truths in these opening chapters of Scripture? My answers to both these questions are a firm "Yes!" and "Yes!"

If we again look at church history, or if we ask those who embrace one of the three Christian views of origins today (young earth creation, progressive creation, evolutionary creation; see Appendix 1), we will discover that there are five main spiritual truths that all believers have always drawn from the biblical accounts of creation: (1) God created the universe and life, (2) the universe and life are very good, (3) only humans are created in the Image of God, (4) only humans are sinful, and (5) God judges humans for their sins. Christians may have passionate debates over how the Lord created the cosmos and living organisms, or they may argue about the meaning and length of the creation days in Genesis 1, but if we ask them to outline the most significant messages of faith in Genesis 1-3, they agree on these five inerrant spiritual truths.

And I know this personally. Over the course of my forty years as a born-again Christian, I have held at different times all three Christian origins positions. But despite my various understandings of God's creative method, the five central messages in Genesis 1-3 have always been foundational to my faith.

In fact, it was during a period in my life when I was seeking God with an open heart that these inerrant spiritual truths led me away from atheism. It was not the interpretation of the length of the creation days in Genesis 1 that changed my immoral lifestyle. Instead, it was the mystical "voice" of God in Scripture that convicted me that there really was a Creator and that I was a sinner. And it was not how God created the world that led me to the Lord, but rather the words of Jesus to the woman caught in adultery, "Go now and leave your life of sin" (Jn. 8:11).

HERMENEUTICAL PRINCIPLE 10

Modern Science & Paraphrase Biblical Translation

How many of us today have actually held a mustard seed in our hand? I haven't. So for most of us, we do not fully appreciate Jesus' parable of the mustard seed within its ancient context. But to assist us, there are modern translators of Scripture who take the inerrant spiritual truths and recast them using modern ideas that are familiar to us. These Bibles are known as "paraphrase translations." This method of translation opens the way for substituting the ancient science in Scripture with modern scientific concepts in order to deliver the biblical messages of faith to our twenty-first century generation of Christians.

To begin exploring this approach, let's look at Eugene Peterson's paraphrase translation of the parable of the mustard seed in Mark 4:30-32. Jesus asked, "How can we picture God's kingdom? What kind of story can we use? It's like a pine nut. When it lands on the ground it is quite small as seeds go, yet once it is planted it grows into a huge pine tree with thick branches. Eagles nest in it."[1]

Of course, if we examine Mark 4:30-32 in the original Greek language of the New Testament, there is no reference to a pine nut, a pine tree, or eagles. Yet I believe that Peterson has made this parable much more relevant and understandable for us today. Instead of using the incidental vessel of a mustard seed to explain the kingdom of God, Peterson employs a pine nut, which is well-known to most Americans. In other words, he exchanges the ancient scientific idea that the mustard seed is the smallest of all seeds on earth with an idea from modern botany that most of us are aware of. Do you think he is successful? I do.

Peterson then continues his updated translation in Mark 4:33-34. "With many stories like these, Jesus presented his *message* to his disciples, fitting the stories to their experience and maturity. He was never without a story when he spoke. When he was alone with his disciples, he went over everything, sorting out the tangles, and untying the knots."[2] Peterson echoes the Message-Incident Principle and the principle of accommodation. The purpose of Jesus' stories was to reveal a "message" of faith and the Lord did so by "fitting" or accommodating these spiritual truths to a level of "experience and maturity" that his disciples could understand.

In fact, the title of Peterson's paraphrase translation is *The Message New Testament*, which clearly reflects the basic concept behind the Message-Incident Principle. He notes, "This version of the New Testament in a contemporary idiom keeps the language of the Message current and fresh and understandable in the same language in which we do our shopping, talk with our friends, worry about world affairs, and teach our children their table manners."[3] Peterson appeals to Jesus in justifying his method of translation. He argues that the Lord had a "preference for down-to-earth stories" that could be easily understood by "common people." Moreover, Peterson points to the greatest act of accommodation—the Incarnation. "For Jesus is the descent of God to our lives, just as they are, not the ascent of our lives to God."[4]

Peterson also introduces us to a valuable insight regarding the original Greek language that is used in the New Testament.

> In the Greek-speaking world of that day, there were two levels of language: formal and informal. Formal language was used to write philosophy and history, government decrees and epic poetry. If someone were to sit down and consciously write for posterity, it would of course be written in the formal language with its learned vocabulary and precise diction. But if the writing was routine—shopping lists, family letters, bills and receipts—it was written in the common, informal idiom of everyday speech, street language. And this is the language used throughout the New Testament.[5]

It might come as a shock to many Christians, but the Holy Spirit-inspired writers of the New Testament did not use formal Greek. Instead, they use Koine Greek. The Greek adjective *koinos* means "common." Peterson adds this was the "street language of the day" and a "rough and earthy language."[6] For me, this is subtle and powerful evidence that indicates God accommodates and descends to the level of the average man and woman in the streets. The message of faith is for everyone, not just those who enjoy higher education and use formal language. In this way, modern paraphrase translations of Scripture attempt to keep the Word of God in the words of common people today.

Biblical Creation Accounts

In the light of Peterson's paraphrase of the parable of the mustard seed in Mark 4:30-32, I believe that inerrant messages of faith in Genesis 1 can be translated in a similar way by employing a modern scientific view of origins as an incidental vessel. Here is the NIV translation of the first five verses of the Bible:

> [1] In the beginning God created the heavens and the earth. [2] Now the earth was formless and empty, darkness was over the surface of the deep, and the Spirit of God was hovering over the waters. [3] And God said, "Let there be light," and there was light. [4] God saw the light was good, and he separated the light from darkness. [5] God called the light "day," and the darkness he called "night." And there was evening, and there was morning—the first day.

Today, nearly all scientists (98%) accept the evolution of the universe and living organisms.[7] Cosmological evolution asserts that the universe began about 14 billion years ago with a gargantuan explosion called "The Big Bang." Space, time, and matter emerged following this explosive event, and then later, stars (suns), planets, and moons slowly evolved through natural processes. Christians who accept cosmological evolution believe that God initiated the Big Bang and created it out of nothing.

In using cosmological evolution as an incidental modern scientific vessel, we can offer the following updated paraphrase of Genesis 1:1-5.

> [1] Over a period of billions of years, God created the heavens and the earth through an evolutionary process. [2] Now the world did not exist before the creation of space, time, and matter. [3] And God said, "Let there be a gargantuan explosion." And there was a gargantuan explosion. [4] God saw it was good, and he separated the explosion from nothingness. [5] God called the explosion "The Big Bang." This was the first stage of cosmological evolution.

To be sure, some Christians might be uncomfortable with this modernized and accommodated version of the opening verses of the Bible. However, it firmly maintains the two central messages of faith in this passage: (1) God is the Creator of the world, and (2) the creation is good. In this way, by using cosmological evolution as an incidental vessel, we can deliver these inerrant biblical truths to our modern scientific generation.

I also believe that a paraphrase translation of the creation of land animals and humans in Genesis 1:24-27 is possible by employing the modern science of biological evolution. The NIV Bible records:

> [24] And God said, "Let the land produce living creatures according to their kinds: the livestock, the creatures that move along the ground, and the wild animals, each according to its kind." And it was so. [25] God made the wild animals according to their kinds, the livestock according to their kinds, and all the creatures that move along the ground according to their kinds. And God saw that it was good.
>
> [26] Then God said, "Let us make mankind in our image, in our likeness, so that they may rule over the fish in the sea and the birds in the sky, over the livestock and all the wild animals, and over all the creatures that move along the ground." [27] So God created mankind in his own image, in the image of God he created them; male and female he created them.

The theory of biological evolution asserts that land animals evolved from fish. Amphibians were the first animals to come onto dry ground, then reptiles evolved from them, and reptiles later evolved into mammals, including humans, who appeared at the very end of the evolutionary process. Here is a modernized paraphrase of Genesis 1:24-27 that uses biological evolution as an incidental vessel to transport inerrant spiritual truths.

> 24 And God said, "Let living creatures evolve into different species: fish evolving into different species of amphibians, amphibians evolving into different species of reptiles, and reptiles evolving into different species of mammals." And it was so. 25 God evolved different species of amphibians, different species of reptiles, and different species of mammals. And God saw that it was good.
>
> 26 Then God said, "Let us use an evolutionary process to make mankind in our image, in our likeness, so that they may rule over the fish in the sea and the birds in the sky, and over the amphibians, reptiles, and mammals that move along the ground. 27 So from pre-human creatures God evolved mankind in his own image, in the image of God he created them; male and female he created them through evolution.

Again, many Christians might find this updated translation quite disturbing. I apologize if this is the case for you. Yet the eternal messages of Genesis 1 are preserved: (1) God created land animals and humans, (2) God declares the creation of land animals is good, (3) only humans are created in the Image of God, and (4) men and women are placed as rulers over all the fish, birds, and land animals.

The scientifically modernized paraphrase translations of Genesis 1:1-5 and 1:24-27 presented above can be helpful to people who accept cosmological and biological evolution. Instead of stumbling over Genesis 1 and the ancient science of the *de novo* creation of the universe and life, as found in standard Bible translations, these updated renditions focus readers on the inerrant messages of faith. In particular, paraphrase translations direct our attention to *who* created the world—the God of the Bible.

HERMENEUTICAL PRINCIPLE 11

Textual Criticism

Regrettably, the word "criticism" usually carries a negative meaning for many people today. But in biblical hermeneutics, this term simply refers to the careful analysis of Scripture using different methods. One form of biblical criticism is textual criticism. In the same way that there are many different types of medical doctors (psychiatrists, ophthalmologists, cardiologists, etc.), there are also various kinds of theologians. Textual critics specialize in examining ancient manuscripts of the Bible and ancient commentaries that cite and refer to scriptural passages.

For example, with regard to the New Testament, there are over five thousand ancient documents, and some variations exist between them. The task of textual criticism is to establish the original biblical text because it no longer exists. Sometimes this analysis may lead to more than one version of a verse, and these variants often appear in the footnotes of Bibles. It is important to underline that though there are textual debates over some verses in Scripture, the modern translations of the Word of God are very good and very reliable. Read any English translation on your knees with an open heart and you will not only discern the central messages of the Christian Faith, but you will also personally meet Jesus.

Of course, most people today have not studied the original biblical languages to the level of expertise of a textual critic. Yet by using some standard interpretive tools, it is possible for Christians to look beyond the English translations to the original ancient passages of Scripture. To illustrate what I mean, take the translation of the mustard seed parable in the 1978 New International Version (NIV) of the Bible. In Mark 4:30-

THE BIBLE & ANCIENT SCIENCE

31, Jesus asks, "What shall we say the Kingdom of God is like, or what parable shall we use to describe it? It is like a mustard seed, which is the smallest seed *you plant* in the ground" (my italics).

However, if you will recall from Hermeneutical Principle 7, I used the 2011 NIV and verse 31 was translated, "It is like a mustard seed, which is the smallest of all the seeds on the earth." Did 2011 NIV translators leave out the words "you plant," or did the 1978 NIV translators add them to this verse? Anyone examining an interlinear Greek-English New Testament can find the answer:[1]

micro		pan	
μικροτερον	ον	παντων	των
smaller	being	all	the

sperma	epi		gē
σπερματων	επι	της	γης
seeds	on	the	earth

It is clear that the words "you plant" are not in the original Greek. The personal pronoun "you" in either the singular (συ) or plural (ὑμεις) is missing, and so too the Greek verb "plant" (φυτευω). We can also examine the variants of the Greek New Testament manuscripts to see whether "you plant" is found in some ancient manuscripts. Once again, the words "you" and "plant" do not appear.[2] In a footnote regarding Mark 4:31 in the Greek New Testament, we find:

ως . . .	σιναπεως
as	mustard seed

What are we to make of the addition of these two words in the 1978 NIV translation? First, I need to emphasize that the translators of the NIV are some of the very best biblical scholars in the world and their mastery of the ancient Greek is second to none. Be assured this was *not* a mistake on their part. Second, I suspect that they added "you plant" to restrict the number of seeds to only those that Jewish people planted in their fields, and not to all the seeds that exist on earth. In this way, these translators help modern readers avoid an unnecessary conflict between Scripture and science should someone be aware that there are seeds much smaller than the mustard seed, such as orchid seeds. In other words, the 1978 NIV translation of Mark 4:31 is an accommodation to our generation.

Biblical Creation Accounts

It may come as a surprise to many Christians that there is a textual debate over the translation of the very first verse in the Bible, and how to arrange the first three verses. Let's look at four different well-known translations of Genesis 1:1-3.

King James Version (1611)

> [1] In the beginning God created the heaven and the earth.
> [2] And the earth was without form, and void; and darkness was upon the face of the deep. And the Spirit of God moved upon the face of the waters. [3] And God said, Let there be light: and there was light.
>
> Footnote: none

This famous and beloved English translation of the Bible has served the church for over four hundred years. I became a Christian reading this version of the Gospel of John. The KJV contends that the first three verses of Genesis present a sequential order of events. First, God creates the heaven and earth in verse 1, then the Spirit of God moves over the formless and void dark watery earth in verse 2, and finally God creates light in verse 3.

Jewish Bible Society (1970)

> [1] When God began to create the heaven and the earth—
> [2] the earth being unformed and void, with darkness over the surface of the deep and a wind from God sweeping over the water— [3] God said, "Let there be light," and there was light.
>
> Footnote: or In the beginning God created

In contrast to the traditional translation of the opening words of Scripture "In the beginning God created . . . ," this Jewish Bible translates "When God began to create . . ." Notably, verse 2 functions like a parenthesis and this scene-setting sentence describes the world before God's creative action, which begins with the creation of light in verse 3.

The JBS translation also has a footnote suggesting the traditional translation of Genesis 1:1 is possible. This is evidence that there is a textual debate among biblical scholars regarding this first verse.

New Revised Standard Version (1991)

> ¹ In the beginning when God created the heavens and the earth, ² the earth was a formless void and darkness covered the face of the deep, while a wind from God swept over the face of the waters. ³ Then God said, "Let there be light"; and there was light.
>
> > Footnote: or When God began to create or In the beginning God created; v.2: or while the spirit of God or while a mighty wind

The NRSV 1991 is often considered to be the best modern translation of the Bible. The various renditions presented in the footnote are more evidence that textual scholars continue to debate the translation of these verses. Verse 1 seems to be a combination of the traditional "In the beginning God . . . " and the modern "When God began to create . . ." Similar to the JBS translation, Genesis 1:2 is the opening scene of this creation account, and Genesis 1:3 presents God's first act of creation with the origin of light.

New International Version (2011)

> ¹ In the beginning God created the heavens and the earth.
> ² Now the earth was formless and empty, darkness was over the surface of the deep, and the Spirit of God was hovering over the waters. ³ And God said, "Let there be light," and there was light.
>
> > Footnote: none

This modern evangelical translation of Genesis 1:1-3 suggests that the first verse of the Bible is a summary statement or a title of the Genesis 1 creation account. Like the JBS and NRSV, Genesis 1:2 is the opening scene of the account with the Spirit of God hovering over the dark, watery, formless, and empty earth. The adverb "Now" at the beginning of verse 2 is used to emphasize that this is the scene-setting sentence. God's creative action then begins in verse 3 with the creation of light.

To be sure, some Christians are uncomfortable when they first discover that we are not completely certain how to translate the first verse of Scripture, or how to arrange the first three verses. I was shocked when a professor introduced me to this issue in seminary. Like most Christians, I assumed that God would make the very first verse of the Bible absolutely clear. But that was my mistaken assumption. This textual debate led me to make two valuable conclusions.

First, we should never lose perspective. All of the four translations of Genesis 1:1 above reveal the same central message of faith—the God of the Bible is the Creator of the heavens and the earth. This is the power of the Word of God. Despite textual debates and the limitations of biblical translations, the Lord ensures that his inerrant spiritual truths are always revealed to those seeking him. This is another affirmation of the sufficiency of Scripture and the proficiency of its readers, as we noted in Hermeneutical Principle 9.

Second, I came to fully realise that biblical interpretation always begins with *submitting to the very words in the Word of God*. We may not like the idea that we do not know with absolute certainty how to translate the first verse of the Bible, but this is the biblical evidence that exists so far. And even if we are troubled by this evidence, we need to accept it despite our assumptions and expectations about how we think the Holy Spirit is supposed to have revealed in Scripture.

The debate over the translation of Genesis 1:1 is an excellent example of the importance of textual criticism. We do not have the original Hebrew text of this creation account. In fact, ancient Hebrew writing only had consonants and did not have vowels. Like the Hebrew used in Israel today, readers supplied the vowels as they read. Consequently, textual critics recognize that there are two acceptable ways to add the vowels, and this is why there are two basic translations of the first verse of the Bible: "In the beginning God created..." and "When God began to create..."

But more significantly, modern translations like the JBS, NRSV, and NIV have Genesis 1:2 as the opening scene in the Genesis 1 account of creation. In biblical Hebrew, sentences that are sequential in

THE BIBLE & ANCIENT SCIENCE

time typically have a verb at the beginning of the sentence. However, Genesis 1:2 has a noun and indicates a sequential break from Genesis 1:1.[3] Therefore, the opening scene in the Bible describes a dark, watery, formless, and empty earth already in existence before God starts his creative activity on the first day of creation. This is known as the "pre-creative state."

Learning about the pre-creative state in Genesis 1:2 was another shocking experience in my theological education. In fact, it was a crushing moment for me. I was a young earth creationist at the time, and reading the second verse of the Bible in the Hebrew language made it clear that the earth was already in existence *before* God began his creative activity in the Genesis 1 account of creation. In fact, there was no mention of when the earth was created, or whether if in fact it was even created.[4] The implications of this biblical evidence were massive. As a young earth creationist, I could no longer date the age of the earth using Scripture, because the Bible never states when the earth was created.

This led to another troubling question in my mind. Does the pre-creative state in Genesis 1:2 challenge the traditional Christian doctrine that God created everything out of nothing (Latin: *creatio ex nihilo*)? But as I became more aware of the ancient science in the Bible, this problem was quickly resolved. Belief in a pre-creative state was the science-of-the-day when Genesis 1:2 was composed. Therefore, by applying the Message-Incident Principle, the pre-creative state is an incidental vessel like the 3-tier universe. It delivers the inerrant spiritual truth that God is in total and complete control of the world, even from its very beginning.

There was one last hermeneutical lesson that I learned from Genesis 1:2 during my time in seminary. Biblical revelation is *progressive* and finds its *fulfillment* in Jesus Christ. In Matthew 5:17 the Lord states, "Do not think that I have come to abolish the Law or the Prophets [i.e., the Old Testament]; I have not come to abolish them but to fulfill them." For example, with regard to making amends (atonement) for our sins, Christians today do not sacrifice animals over and over again as commanded in the Old Testament Book of Leviticus in chapters 4-17. Instead, Hebrews 10:10 states, "We have been made holy through the sacrifice of the body of Jesus Christ *once for all*" (my italics).

Similarly, in order to deal with the doctrine of creation-out-of-nothing, we need to turn to the New Testament, and not to Genesis 1:2. Hebrews 11:3 asserts, "By faith we understand that the universe was formed at God's command, so that what is seen was not made out of what was visible." In particular, as Christians we must focus on Jesus. Colossians 1:16-17 reveals, "For in him [Jesus] all things were created: things in heaven and on earth, visible and invisible, whether thrones or powers or rulers or authorities; all things have been created through him and for him. *He is before all things*, and in him all things hold together" (my italics). By acknowledging that biblical revelation is progressive, it became clear to me that Jesus did indeed create the world out of nothing. The Christian doctrine of *creatio ex nihilo* is true.

HERMENEUTICAL PRINCIPLE 12

Implicit Scientific Concepts

Previously I introduced the concept of hermeneutical horizons between the Bible and modern readers by using the example of my students attempting to interpret and draw the dark, watery, formless, and empty earth in Genesis 1:2. Most of them assumed the word "earth" in this verse refers to our spherical planet (Fig. 4-1, p. 33). By failing to recognize and respect the ancient science in Scripture, they eisegetically forced their twenty-first century understanding of the structure of the earth into an ancient text that actually includes a 3-tier universe and a flat earth.

Implicit scientific concepts refer to scientific ideas-of-the-day that are assumed by both the biblical writers and the modern readers of Scripture. Many times, these notions about the natural world are deeply embedded in the mind and often function at a tacit level. The Latin adjective *tacitus* means "silent." If asked directly about an implicit scientific concept, a biblical writer or a modern reader could explain their view. Regarding the shape of the earth, the former would say it is flat, while the latter would claim it is spherical. But often these concepts are simply assumed and not expressed explicitly.

As a result, modern readers usually filter the word "earth" in the Bible through their modern scientific mindset and take for granted a spherical planet when in fact the biblical writers implicitly understood an earth that was flat. Not being fully aware of our implicit scientific concepts and those of the inspired biblical writers usually results in mistaken interpretations of Scripture. As I jokingly say to my students, implicit scientific concepts could also be termed the "hermeneutical silent killers!"

IMPLICIT SCIENTIFIC CONCEPTS

To further explain this principle of interpretation, let's look at another ancient science in the Bible—passages that deal with human reproductive biology. In order to understand what ancient people believed and assumed, we need to think like them. For example, during sexual intercourse they would have experienced that men ejaculate, and women do not. Since agriculture played a huge role in their lives, it was natural for ancient people to perceive a similarity between plant seeds and the "seeds" ejaculated by a man. It is important to note that it is a common practice in both ancient and modern science to employ familiar objects and processes as models and metaphors to describe and to explain the natural world.[1] Consequently, throughout the Bible only men are said to have reproductive seed, never women. Let's look at some passages that deal with human reproduction in Scripture.

In the Old Testament, the Hebrew noun *zera'* refers to both the "seeds of plants" and the "reproductive seeds of males."[2] It is related to the verb *zāra'* which means "to seed" or "to sow." In the context of planting crops, a literal translation of Genesis 47:23 records, "Here is seed [*zera'*] so that you can seed [*zāra'*] the ground." Within a sexual context, the phrase "a woman shall be seeded [*zāra'*] with seed [*zera'*]" appears in Numbers 5:28 to indicate she will be inseminated by a man. Similarly, Leviticus 15:32 and 22:4 use the literal expression the "laying down of seed [*zera'*]" or "flow of seed [*zera'*]" to mean the ejaculation of semen.

In the New Testament, the Greek noun *sperma* refers to both "plant seed" and "male sexual seed." In an agricultural context, Jesus began the parable of the weeds in Matthew 13:24, "The kingdom of heaven is like a man who sowed good seed [*sperma*] in his field." The great biblical chapter on faith in Hebrews 11 provides an example of *sperma* in a sexual context with ejaculation. In verse 11 most English Bibles employ a euphemism. The 1978 NIV Bible translates, "By faith Abraham, even though he was past age—and Sarah herself was barren—was enabled to become a father because he considered him [God] faithful who had made the promise."

83

The original Greek for the phrase translated "enabled to become a father" is *katabolēn spermatos*. As we noted earlier in Philippians 2:10, "*kata*" is the preposition "down." The verb *bolē* means "to throw." In the ancient world, *katabolē sperma* was a technical term for ejaculation, and it was used only for men. Therefore, Hebrews 11:11 is best translated, "By faith Abraham, even though he was past age—and Sarah herself was barren—was enabled *to throw down seed* because he considered him [God] faithful who had made the promise." In other words, God miraculously gave Abraham the ability to ejaculate. In doing so, Isaac was born (Gen. 21:2-3) and the Lord fulfilled his promise to Abraham, "I will establish my covenant with him [Isaac] as an everlasting covenant for his descendants after him" (Gen. 17:19).

In light of this ancient understanding of human reproductive biology, it is no co-incidence that children are referred to in agricultural terms as the "fruit of the womb." For instance Psalm 127:3 states, "Sons are a heritage from the Lord, the fruit of the womb a reward" (NRSV). This agricultural expression is even used in referring to Jesus. In Luke 1:42, Elizabeth cries out to Mary who was pregnant with the Lord, "Blessed are you among women, and blessed is the fruit of your womb" (NRSV).

This ancient view of human reproduction is known as "preformatism" or the "1-seed model." Ancient people believed that within each male sexual seed there was a tightly-packed miniscule human. Not having the advantage of microscopes, this was a perfectly logical idea from an ancient phenomenological perspective. Men ejaculate during sex, giving the impression that they are the only contributors of seed in the creation of a human being; while women appear to be only receptacles and nurturers of the seed of males, similar to an agricultural field.[3]

Figure 12-1 presents the hermeneutical horizons between the Bible and the modern reader with regard to human reproductive biology. Of course, it is very tempting for us today to read the Greek word *sperma* in Scripture and assume that it refers to a sperm cell. To be sure, our modern biological term "sperm" does derive from the word *sperma*. However, we must remember that this ancient term referred to a seed that contained a miniature human inside it, as shown in the diagram.

IMPLICIT SCIENTIFIC CONCEPTS

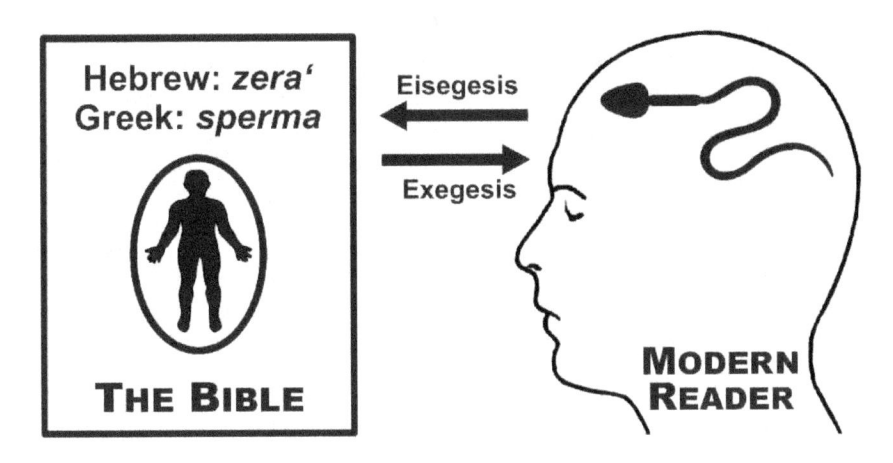

Figure 12-1. Hermeneutical Horizons & Human Reproductive Biology

As modern readers of Scripture, it is easy for us to miss the implicit ancient scientific concepts in the Bible. Take for example Hebrews 7:9-10. "One might say even that Levi, who collects the tenth, paid the tenth through Abraham, because when Melchizedek met Abraham, Levi was still in the body of his ancestor." We know from Genesis 14:18-20 that Abraham gave a tenth of his goods to Melchizedek, who is referred to as "the priest of the most high God." We also know Abraham's descendants. He fathered Isaac (Gen. 21:1-7), who fathered Jacob (Gen. 25:26), who fathered Levi (Gen. 29:34).

From the perspective of modern reproductive biology, it is impossible for Levi to be "in the body" of Abraham because Levi's genes had not yet been made. Three processes of meiosis and three processes of fertilization had to occur before arriving at the fertilized egg that became Levi.

However, Hebrew 7:9-10 makes perfect sense if we recognize and respect the implicit scientific concepts of the biblical writer. According to the ancient reproductive biology of preformatism, when Abraham met Melchizedek, Levi was in the body of his father Jacob, who was in the body of his father Isaac, who was in the body of his father Abraham. In other words, ancient people viewed human reproduction as something similar to a set of Russian dolls with one person inside the other.

THE BIBLE & ANCIENT SCIENCE

In fact, the more accurate translation of the Greek word *osphus* in verse 10 is not "body," but "reproductive organ." Translators again used a euphemism. Therefore, from the perspective of ancient human reproductive biology, the miniature seed of Levi was assumed to be in the reproductive organ of Abraham at the time Abraham met Melchizedek.[4]

The powerful influence of implicit scientific concepts on thinking can be seen in this next example. The ancient reproductive notion of preformatism lasted until the eighteenth century, even when scientists began to examine human sperm through the first microscopes.[5] As Figure 12-2 shows, they imagined an entire minuscule human was folded up in the head of the sperm cell.[6] This demonstrates the impact of implicit scientific assumptions on the ability to interpret scientific evidence. Similar to biblical interpretation, scientists can commit eisegesis by reading foreign and incorrect ideas into their observations of nature. In this case, scientific investigators forced preformatism into their understanding of the structure of a sperm cell.

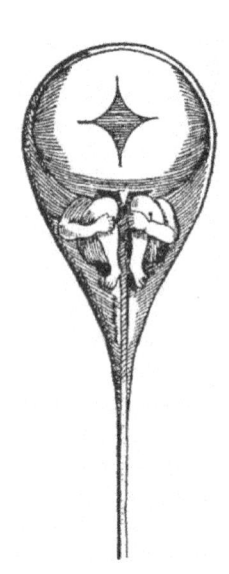

Figure 12-2. Male Seed according to Preformatism

Biblical Creation Accounts

Did the inspired ancient author of Genesis 1 have ingrained implicit scientific concepts when he wrote this account of origins? And do the implicit scientific concepts of modern readers influence their reading of this opening chapter of the Word of God? To answer these questions, let's return to Genesis 1:2 and the pre-creative state. "Now the earth was formless and empty, darkness was over the surface of the deep, and the Spirit of God was hovering over the waters."

As we noted, Hugh Ross is the most important progressive creationist in the world today. He is an astronomer and the president of the Christian organization Reasons to Believe. In his interpretation of the second verse of the Bible, Ross claims, "The events of the six creation

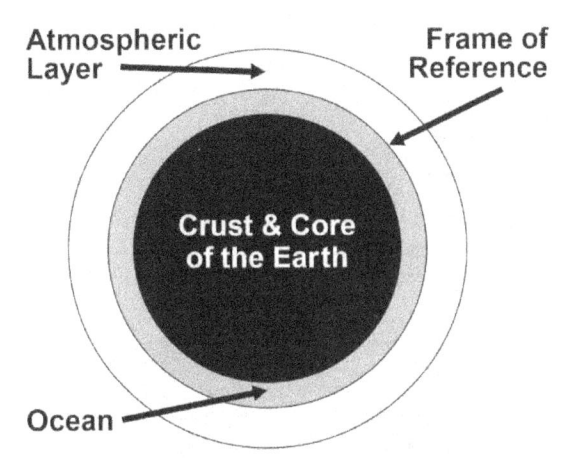

Figure 12-3. Hugh Ross' Depiction of Genesis 1:2

days are described from the vantage point of Earth's primordial (water-covered) surface, underneath the cloud layer, as Genesis 1:2 indicates."[7] To assist his readers to picture the scene in this verse, Ross offers a diagram as redrawn in Figure 12-3.[8]

In continuing his interpretation of Genesis 1:2, Hugh Ross argues, "The observer's vantage point is clearly identified as 'the surface of the deep . . . over the waters.' Yet the vast majority of Genesis commentaries mistakenly proceed as if it were still high in the heavens, somewhere in the starry realm above Earth." Ross concludes, "Obviously, no author writing more than 3,400 years ago, as Moses did, could have so accurately described and sequenced these events, plus the initial conditions, without divine assistance . . . [This is] powerful evidence of the scientific soundness of the Bible . . . the book of Genesis must be supernaturally inspired."[9]

However, I am certain that you have spotted the critical error in Hugh Ross' interpretation of Genesis 1:2. Not only is he a scientific concordist, but his depiction of this verse is dictated by his implicit scientific concept that the earth is a *sphere*. To be more specific, by not recognizing and respecting the ancient understanding of a flat earth in the Bible, Ross eisegetically forces his twenty-first century science into

THE BIBLE & ANCIENT SCIENCE

[Figure: Diagram showing DARKNESS and SPIRIT OF GOD above THE WATERS, with FORMLESS & EMPTY EARTH and THE DEEP below.]

Figure 12-4. Pre-Creative State in Genesis 1:2

Genesis 1:2. In other words, he unwittingly thrusts a foreign idea into the Word of God.

In attempting to picture the scene in Genesis 1:2, we need to think like an ancient person and remember that they looked at nature from their ancient phenomenological perspective. They understood the structure of the world to be a 3-tier universe with a flat earth when the Creator finally completed it. In other words, this was the implicit scientific concept that shaped the thinking of the author of Genesis 1. Consequently, in depicting the pre-creative state in Genesis 1:2, this author would have envisioned a primordial world somewhat similar to Figure 12-4. Darkness was over the flat surface of the waters and the deep since light had yet to be created on the first day of creation (Gen. 1:3). And this body of water covered the formless and empty earth because dry land only appears on day three (Gen. 1:9-10).

But more importantly, the purpose of the Genesis 1 creation account is not to disclose scientific facts about the origin of the world ahead of time, as suggested by scientific concordist Hugh Ross. The message of faith in Genesis 1:2 reveals that the Spirit of God was in total control of the creation from its earliest moments. In order to deliver this inerrant spiritual truth, the Lord accommodated and came down to the intellectual level of the inspired biblical writer and used his implicit scientific concept that the world was a dark and watery place before God's acts of creation in Genesis 1. As a result, the pre-creative state in Genesis 1:2 is an incidental ancient vessel that delivers an inerrant message of faith.

Hugh Ross' interpretation of Genesis 1:2 is eisegetical. But it offers us two valuable hermeneutical insights. First, it is quite easy to overlook the implicit ancient scientific concepts of nature in Scripture, like the pre-creative state. Second, it is very common for Christians to inadvertently force their modern scientific assumptions into the Word of God, such as a spherical earth.

To avoid these hermeneutical errors, we need to become aware of the science-of-the-day when the Holy Spirit inspired the biblical authors. More specifically, it is necessary that we familiarize ourselves with the ancient science of the nations that surrounded the Hebrews and early Christians. In particular, the interpretation of Genesis 1 and 2 requires that we explore creation accounts from ancient Near Eastern (ANE) peoples like the Mesopotamians and Egyptians.*And at all times, we must be careful never to eisegetically impose our modern science into the Word of God.

---

* We will further explore this notion in Hermeneutical Principle 14: Historical Criticism.

HERMENEUTICAL PRINCIPLE 13

Scope of Cognitive Competence

In most of our hermeneutical principles we have been noting that the biblical writers looked at the physical world and thought about it much differently than we do today. To use an optical (visual) scope on a scientific instrument as a metaphor, we could say that everyone views nature and thinks about it through a "cognitive scope." The Latin verb *cognitō* means "to learn" and "acquire knowledge," and the English word "cognition" refers to "the mental processes involved in perceiving, reasoning, and knowing." Within the context of this hermeneutical principle, the term "cognitive scope" depicts the mental or intellectual tools through which everyone sees and understands the natural world.

An implication of the scope of cognition is that our perception and knowledge of nature have limits and boundaries. This is similar to the margins of a visual field when using the optical scope of a scientific instrument. In addition, intellectual tools are profoundly shaped by a person's intelligence, education, culture, and point in history. Therefore, cognitive competence is dependent on how far our mental tools can penetrate the natural world. To illustrate, a magnifying glass is capable of increasing the size of words on the surface of a written page. But it is not competent to view the various molecular structures within the ink or paper of such a page.

Thanks to telescopes and microscopes our modern scope of cognitive competence is much wider and greater than that of ancient people. We can see further out into space and deeper into the cells of our body than any generation before us. This is not to disrespect these individuals from the past, such as the ancient Hebrews and early Christians. Although it is to say that their ability to view and understand nature was significantly less than ours today. This fact has important implications for biblical interpretation. For example, ancient people did not have microscopes, and therefore they did not have the cognitive competence to know that women contributed an egg to the creation of a child.

Let's continue examining the topic of human reproductive biology in Scripture. To begin with a question, "Have you ever noticed in the Bible who is responsible when there are difficulties in conceiving a child?" Yes, it's always the woman. Here are three verses from the Old Testament (NRSV): "Now Sarai was barren; she had no child" (Gen. 11:30), "Isaac prayed to the Lord for his wife [Rebekah], because she was barren" (Gen. 25:21), and "Rachel was barren" (Gen. 29:31).

The translators of these verses use the word "barren" and echo the notion of a field that is desolate and unfertile. Since ancient people understood human sexual reproduction through their experience of agriculture and the sowing of seeds in the ground, it made sense to think that a woman who could not have children was like an unproductive barren field. In fact, the Hebrew word translated as "barren" in these three verses reflects ancient agriculture. It is the adjective *'āqār* and derives from the verb *'āqar*, which means "to uproot" and "pluck up." This verb appears in Ecclesiastes 3:1-2 in an agricultural context. "There is a time for everything . . . a time to plant and a time to uproot [*'āqar*]."

Why would a woman who could not have children be seen as "uprooted" and "plucked up?" First, it is important to emphasize that there is no intention whatsoever to offend women. Had we lived at a time with no microscopes and a narrow scope of cognitive competence, we would have believed this. Second, as we have noted previously, ancient people thought the womb of a woman was similar to a farmer's field in which the seed (*zera'*) of a man was sown (*zāra'*) within her. But in the case of a barren woman, the fetus begins to develop in her body and then it gets pulled up and torn away. Undoubtedly, ancient people would have seen miscarriages and viewed the developing child as similar to a small sprouting plant uprooted from a field.

THE BIBLE & ANCIENT SCIENCE

An ancient understanding of human reproductive problems also appears in the New Testament. Here are three verses from the NRSV Bible: "But they [Zechariah and Elizabeth] had no children, because Elizabeth was barren" (Lk. 1:7), "For the days are surely coming when they will say, 'Blessed are the barren, and the wombs that never bore, and the breasts that never nursed'" (Lk. 23:29), and "Sarah herself was barren" (Heb. 11:11).

Similar to the Old Testament, the translators use the agricultural term "barren" in order to reflect an ancient understanding of reproductive biology. The actual Greek word in these three verses is the adjective *steira*, which is the feminine form of *steiros*. You are correct in thinking the English term "sterile" derives from it. Notably, *steiros* is related to the adjective *stereos*, which has the meanings "hard," "firm," and "solid." In light of these Greek words, why would a woman who cannot conceive a child be seen as "hard"? Ancient people, with their narrow scope of cognitive competence, understood her womb to be like a field with impenetrable compacted soil. The woman receives "seed" (*sperma*) from the man during intercourse, but like a seed on a hard path it never takes root and develops.

The experience of sowing seed in different fields would also have informed the ancient view of human reproductive difficulties. Today we know that the worst rate of germination of seeds like wheat and barley is about 85%.[1] Therefore, if ancient farmers sowed the poorest seeds in a field next to one that used the best seeds, it is doubtful that they would have perceived a difference in productivity between the two crops. From their ancient phenomenological perspective, they would have assumed that seeds are always able to germinate if the soil was good. To use the previous hermeneutical principle, the idea that seeds are 100% efficient was an implicit scientific concept for ancient peoples.

Farmers in the ancient world also knew that different types of soils in which seeds were sown produced different yields. Jesus' parable of the sower uses this ancient agricultural notion (Matt. 13:3-9). The seeds that land on a path never take root. In rocky places with shallow soil, they spring up quickly, but never thrive. And seeds sown among thorns are choked out. But in good soil, they produce a full crop. Thus, it is the type of soil in a field that is the determining factor for whether or not seeds are productive. This ancient phenomenological perception aligns with Mark 4:28, which we examined earlier in Hermeneutical Principle 7. "*All by itself* the soil produces grain" (my italics).

If we recognize and respect that ancient people had a narrow scope of cognitive competence, it is understandable why they believed that the woman was the problem when she could not bear a child. The reproductive seed of a man was assumed to be 100% efficient like that of plant seeds, and the womb of a woman was comparable to the soil in a field.[2] By perceiving and reasoning through ancient agricultural concepts, if the woman was childless, it was because her womb was not fertile for the man's seed. She was deemed to be either "hard" (*stereos*) like compacted ground on a path or "uprooted" (*'āqar*) and similar to a field with a small seedling torn out of it.

Of course, modern science through microscopes and other scientific instruments has discovered that infertility is not limited to females only. We have advanced intellectual tools and a much wider scope of cognitive competence. Men can also be a source of infertility. Standard figures today indicate that this medical condition is caused 30-40% of the time by males, 30-40% by females, and the remainder by both or an unknown factor.[3] Again, the attribution of barrenness to only women in the Bible is in no way intended to disrespect women. This simply was the best human reproductive science-of-the-day, and had we lived at that time, this would have been our understanding.

But more importantly, the Old and New Testament passages cited above that deal with barren women reveal inerrant spiritual truths. First, God is in total control of the natural world; and second, he fulfills his promises. The miraculous birth of children to barren women in the Bible is evidence of the Lord's power over nature and his faithfulness to create a holy people to serve his purposes.

Biblical Creation Accounts

So far we have noted that the Bible includes an ancient understanding of the structure of the world (3-tier universe with a flat earth), an ancient botany (mustard seed smallest of all seeds), and an ancient view of human reproduction (preformatism). It is only consistent that the inspired writers of the biblical creation accounts would also have had a narrow scope of cognitive competence.

Let's consider two examples. In the ancient world there were no telescopes. When ancient men and women looked up to the sky through their naked eyes, they perceived the heavens to be a solid immovable dome above their heads. But thanks to high powered telescopes made during the twentieth century, we now know that the heavens are expanding with galaxies rapidly moving away from each another. Therefore, the narrow scope of cognitive competence of the biblical author of Genesis 1 led him and his readers to believe the universe was *unchanging* and *static*.

Similarly, ancient people did not have microscopes and other advanced scientific instruments to study the genes of plants and animals. It is only in the last few decades that there have been dramatic advances in biology. Thanks to genetic research, we now know that all living organisms are related. This similarity is like the resemblance between the genes in your family members and relatives. In fact, this scientific evidence is some of the best evidence that supports the evolution of life. However, the biblical authors did not have the privilege of knowing this amazing information. Limited to their naked eyes, they only saw that plants and animals reproduce "after their/its kinds," as stated ten times in Genesis 1. In this way, they reasonably came to the conclusion that living organisms *never changed* and were *static*.

But here is the power of the Word of God. Despite the narrow scope of cognitive competence of the authors of the biblical creation accounts and their static picture of the universe and life, the central inerrant message of faith has been understood by every generation of Christian. The God of the Bible is the Creator of the heavens and the earth and all the living organisms. Indeed, Scripture is sufficient in revealing who made the world, and we are proficient in knowing that he alone is the Creator.

Excursus
"Hermeneutical Brakes"

Learning about the principles of biblical interpretation for the first time can be challenging and even threatening for many Christians. Some assume that these concepts put believers on the so-called "slippery slope" that can lead to a loss of faith. To be sure, hermeneutical principles certainly raise questions about some things we were taught in our churches and Sunday schools. However, in my experience, grasping the notion that the biblical writers had a narrow perception and understanding of the world has strengthened my faith in the Word of God. Instead of sliding down the slippery slope, the scope of cognitive competence functions like a set of "hermeneutical brakes" that stops us from sliding away from the Lord.

Let me explain this idea by asking some questions. Were the biblical writers competent in understanding the shape of the earth? Did they know the size of mustard seeds compared to all the other seeds on earth? Or were ancient authors of Scripture and their readers capable of comprehending the actual cause of human reproductive difficulties? I am certain that you will agree that they were not competent to know these things about the natural world, because they had a narrow scope of cognitive competence. In particular, they did not enjoy marvelous scientific instruments like telescopes or microscopes as we do today.

But now consider Jesus' first miracle at the wedding in Cana, as described in John 2:1-11. After running out of wine at the banquet, the Lord told some servants to fill six large stone jars with water. He then asked them to draw out some of the water that had been turned to wine and to bring it to the master of the banquet. The master tasted the wine and then remarked to the bridegroom, "Everyone brings out the choice wine first and then the cheaper wine after the guests have had too much to drink; but you have saved the best till now" (v. 10).

In light of this passage, we can ask a question. Were the servants who filled the jars with water competent to know that the water had really been changed into wine? In other words, did they have the intellectual tools to perceive and understand that a miracle had indeed happened? Yes, because it was well within their scope of cognitive competence.

Mark 2:1-12 reports another of Jesus' amazing miracles. He heals a paralytic man. Four men brought him on a mat and lowered him through an opening in the roof of someone's home, so that he could meet the Lord. After saying to the man that his sins were forgiven, Jesus confirmed his authority to forgive sins by saying to the paralytic, "I tell you, get up, take your mat and go home" (v. 11). The man then "got up, took his mat and walked out in full view of them" (v. 12).

Was it within the scope of cognitive competence of the paralytic man and everyone around him in the home to know that Jesus had healed him of his paralysis? Yes. They certainly had the intellectual tools to determine whether or not a miracle had really happened. To use their very words, "We have never seen anything like this!" (v. 12).

Consider one last example regarding Jesus. During his day there were numerous crucifixions and most people would have witnessed some of these cruel and barbaric executions. They knew that when a crucified individual had stopped breathing and his body becomes cold and stiff, that person had died. And most importantly, ancient people were fully aware that a dead person does not come back to life again.

However, three days after the death of Jesus on the Cross, many men and women discovered that he was alive and had been resurrected. Matthew 28:9 records that Mary Magdalene and the other Mary met the Lord and that they "clasped his feet and worshipped him." Luke 24:36-39 states that Jesus appeared to his disciples and said to them, "Look at my hands and my feet. It is I myself! Touch and see; a ghost does not have flesh and bones, as you see I have." Later the disciples even gave the Lord "a piece of broiled fish, and he took it and ate it in their presence" (v. 42-43).

We can again ask the question. Was it within the cognitive limits of these ancient people to know that Jesus had physically died and then rose physically from death? My answer is an emphatic "YES!!!" They clasped and touched him and even saw him eat fish right in front of them. As a result, they knew that he was not a ghost or a delusion. These ancient men and women would definitely have been competent to determine whether or not the Lord had been resurrected bodily from the grave.

Let's put all the questions that I have asked above in perspective. Knowing the shape of the earth, the size of seeds, or the cause for reproductive problems is not essential for being a Christian. I am sure you will agree that these issues are completely irrelevant to our faith.

However, the miracles of Jesus and his bodily resurrection after his physical death on the Cross are foundational truths of Christianity. First century men and women would have been well-equipped with the intellectual tools to see and know that these miraculous events had actually taken place. In other words, it was well within their scope of cognitive competence. Therefore, this hermeneutical principle acts like a set of "hermeneutical brakes." It stops anyone from sliding down the so-called "slippery slope" and doubting the testimony recorded in the Bible of those who saw and experienced Jesus' miracles and resurrection. Personally, I believe that all these miraculous events did occur. They are non-negotiable and inerrant truths of my faith and walk with the resurrected and living Lord of my life, Jesus Christ.

HERMENEUTICAL PRINCIPLE 14

Historical Criticism

As we noted previously, the term "criticism" often has a negative connotation today. But it carries a positive meaning in biblical interpretation and refers to the thorough analysis of Scripture using a variety of methods. Historical criticism is a form of biblical criticism. This approach examines the literature and archeological evidence of nations that surrounded ancient Israel and the early church. By studying this historical information, we can gain a more complete picture of the ideas and mindsets of the ancient world, and as a result, a better understanding of the Bible and the setting in which it was written and read.

To appreciate the importance of historical criticism, assume for a moment that two thousand years from now only one book remained about the United States and that it was written by an American author in our generation. If other books from this same period were discovered in Canada and Mexico that referred to America, they undoubtedly would increase our knowledge of the United States and the historical setting of the nation at that time. Similarly, reading the literature and exploring the archeology of ancient nations surrounding the biblical writers—such as the Mesopotamians, Egyptians, Greeks, and Romans—will offer valuable information for interpreting the Word of God.[1]

Let's continue examining how human reproduction was understood in the ancient world and offer an example of historical criticism. In the fifth century BC, the Greek author Aeschylus states, "The mother of what is called her child—is no parent of it, but nurse only of the young life that is sown in her. The parent is the male, and she but a stranger, a friend, who, if fate spares his plant, preserves it till it puts forth."[2] This writer accepted the pagan belief that fate controls the destiny of people.

HISTORICAL CRITICISM

It is evident that Aeschylus accepts the ancient 1-seed model of reproductive biology (preformatism) when he states that the mother of a child is not the parent. I don't think anyone should try saying that to a pregnant woman! His explicit use of the agricultural terms "sown" and "plant" also reflects this ancient science. Aeschylus reveals that preformatism was the science-of-the-day in the ancient world, and we should not be surprised if this view of human reproduction appears in Scripture.

A biblical verse that we saw earlier includes ancient reproductive biology and agricultural terminology similar to that in the Aeschylus quotation. When Mary was pregnant with Jesus, she visited her relative Elizabeth. Filled with the Holy Spirit, Elizabeth cried out, "Blessed are you among women, and blessed is the *fruit of your womb*!" (Lk. 1:42; NRSV, my italics). Elizabeth then continued, "Blessed is she who has believed that the Lord would fulfill his promises to her" (v. 45). In sharp contrast to Aeschylus' pagan belief in fate, the spiritual truth in this biblical passage is that the God of the Bible is in complete control of the world. He promised Mary that she would have a child even though she was a virgin, and through a miracle she became pregnant and gave birth to Jesus. Indeed, our God keeps his promises because "nothing is impossible with God" (Lk. 1:37; NIV 1978).

Biblical scholar George Eldon Ladd offers an insight that captures the central concept behind historical criticism. "The Bible is the Word of God given in the words of men in history."[3] Ladd's aphorism reflects the two components of the Message-Incident Principle, which I have depicted using inclusive language in Figure 14-1. The message of faith is the "Word of God." The incidental ancient vessel that delivers inerrant

Bible
↗ **MESSAGE**
Word of God
INERRANT

↘ **INCIDENT**
Words of Humans
in History

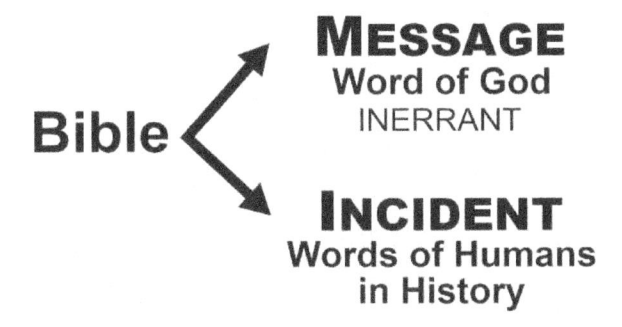

Figure 14-1. G.E. Ladd's Aphorism & Message-Incident Principle

spiritual truths are the "words of humans in history." And these words of men and women include ancient understandings of nature, such as the science-of-the-day.

To be sure, it might be troubling for some Christians to think that the Bible reflects the historical period when it was written. Stated another way, the Word of God is historically conditioned. But it is. Scripture was inspired by the Holy Spirit and written down by real human authors at different times in real history. To use an analogy, the Bible contains a body of eternal spiritual truths that wears ancient historical clothing.

Another aphorism that shows the significance of historical criticism comes from biblical scholar John Walton. He encourages Christians to understand, "The Bible is written *for* us (indeed, for everyone), it is not written *to* us."[4] In other words, Holy Scripture is *for* every man and woman throughout time, but it was directed *to* a specific ancient audience at a specific point in the past. The challenge of biblical interpretation is to identify the historically conditioned elements in Scripture in order to separate them from the eternal spiritual truths. And the best way to identify incidental ancient components in the Word of God is to study the literature and archeology of ancient peoples surrounding the biblical authors.

John Walton also offers a valuable insight regarding the ancient science in Scripture. "Through the entire Bible, there is *not a single instance* in which God revealed to Israel a science beyond their own culture. No passage offers a scientific perspective that was not common to the Old World science of antiquity."[5] Similarly, Walton asserts that "there is *not a single instance* in the Old Testament of God giving scientific information that transcended the understanding of the Israelite audience."[6] It is thanks to John Walton's extensive knowledge of the writings and archeology of ancient nations surrounding the biblical authors that he arrives at this important conclusion. A significant implication of the "Old World science" in Scripture is that it cannot be aligned with modern science. Or stated more incisively, historical criticism undermines scientific concordism.

HISTORICAL CRITICISM

The 3-Tier Universe in Ancient Egypt & Ancient Mesopotamia

The study of historical criticism also offers insights into how ancient people surrounding the biblical writers understood the structure and operation of the world. In particular, the ancient Egyptians and ancient Mesopotamians believed that the universe had basically three levels—the heavens above, the surface of the earth in the middle, and the underworld below. As we have noted before, this conception of the cosmos is often termed the "3-tier universe." In contrast to our modern phenomenological perspective of nature, these ancient individuals viewed and understood the physical world from an ancient phenomenological perspective, as distinguished in Figure 14-2.

The idea that the world was made of three levels is quite reasonable because ancient people had a narrow scope of cognitive competence and did not enjoy modern scientific instruments like telescopes. The dome of the sky and the circumference of the horizon give the impression that there is a firm immovable structure overhead, similar to an inverted bowl. In the ancient Near East, this dome was called the "firmament." The blue of the sky also led to the belief a heavenly sea was above a solid transparent firmament. Moreover, the earth appears to be flat when viewed from an elevated place like a mountain. And the daily movement of the sun across the sky, dipping below the horizon in the west and later rising in the east, convinced ancient people that the sun literally traveled through a region that is under the earth.

ANCIENT
Unaided
Physical Senses
LITERAL & ACTUAL
Dome of Heaven
Flat Earth
Underworld

MODERN
Aided by
Scientific Instruments
APPEARANCE OF
"Dome" of Heaven
"Flat" Earth
"Under" the Horizon

◄─────────────►

PHENOMENOLOGICAL PERSPECTIVES

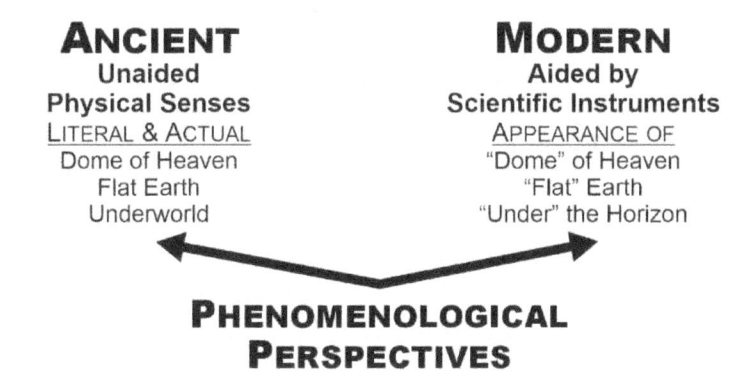

Figure 14-2. Ancient & Modern Phenomenological Perspectives

THE BIBLE & ANCIENT SCIENCE

Figure 14-3. Ancient Egyptian 3-Tier Universe

Of course, the ancient Egyptians and ancient Mesopotamians had no way of knowing that the dome of the sky and its blue color is only a visual effect caused by the scattering of shortwave light in the upper atmosphere. Nor were they aware that the earth is a spherical planet, or that the earth rotated on its axis giving the appearance of the sun dropping under the horizon in the evening and rising in the morning. But again, these ancient scientific ideas were perfectly logical considering ancient people perceived nature from an ancient phenomenological perspective.

Now equipped with an appreciation of the mindset of ancient people, we can examine diagrams and texts from ancient Egypt and ancient Mesopotamia and view the physical world through their ancient eyes.

Figure 14-3 presents an image of the world found on an Egyptian coffin dated between 1570 and 1085 BC.[7] The Egyptians believed that the firmament (shaded) was speckled with the stars and that it was the goddess Nut. Below the firmament, with arms raised, is the god of the air Shu. At the horizon, the ends of the firmament come to the base of an extended level surface. The earth god Geb is reclined, indicating that the earth is basically flat. The sun god Re with a sun-disk on his head is in a boat, which demonstrates the Egyptians believed in a heavenly sea above the firmament. The two images of Re's boat affirms the belief that the sun literally moves across the sky. In the lower right corner of the diagram

HISTORICAL CRITICISM

Figure 14-4. Ancient Egyptian Geography

at the western horizon, the god of the underworld Osiris is ready to receive the sun god Re, who then travels through the underworld to rise again in the east.

Figure 14-4 is a diagram of the earth that appears on a fourth-century BC coffin from Egypt.[8] Note that this is the entire earth according to the ancient Egyptians. They also believed that the earth is flat and circular, and that it is surrounded by a circumferential sea (shaded). Egypt is in the middle of the world, inside the thick black circle drawn on the diagram. The region between the Egyptian border and the shoreline of the sea are foreign lands. And the figures in black with the boat of the sun god Re near their raised arm are the eastern (left) and western (right) doorways of the underworld.

Ancient texts from Egypt also mention this ancient geography. An inscription referring to the Pharaoh Queen Hatsheput (1478-1458 BC) states that "the lands were hers, the countries were hers, all that the heavens cover [like an inverted bowl], all that the sea encircles."[9] Similarly, a victory song to Thutmose III (1458-1425 BC) asserts that he defeated "the ends of the lands; that which the ocean encircles."[10] And an inscription from the period of Ramses III (1186-1155 BC) even men-

103

THE BIBLE & ANCIENT SCIENCE

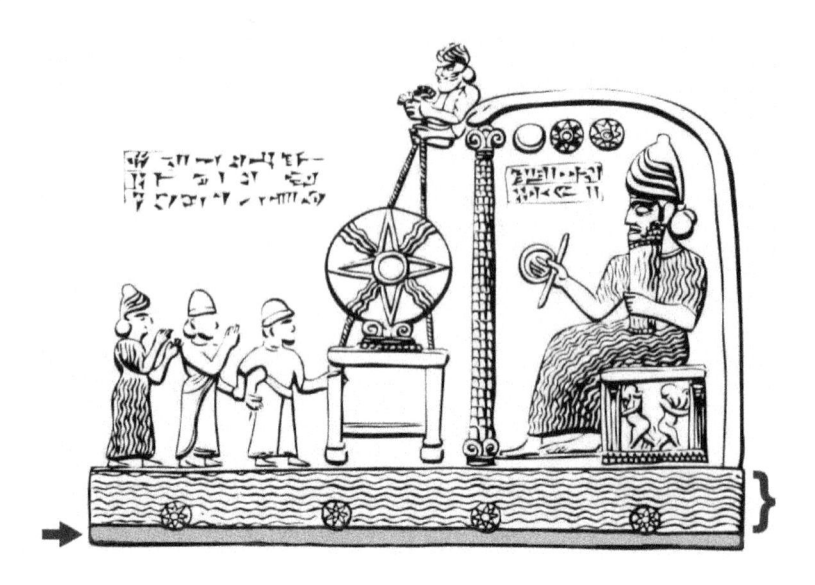

Figure 14-5. Ancient Mesopotamian Astronomy

tions specifically the "circle of the earth,"[11] indicating the earth was shaped like a flat disc. In addition, the ancient Egyptians believed that the earth floated upon a deep primeval body of water that they called "Nun." During the reign of Ramses III, a hymn praises the creator god Ptah, "who founded the earth . . . who surrounded it with Nun, and the sea."[12]

Figure 14-5 depicts a Mesopotamian understanding of the structure of heaven, found on the Shamash Tablet dated around 850 BC.[13] The sun god Shamash is in heaven and he sits on his throne inside a shrine. A large solar disk rests on an altar, and to the left the King of Babylon is led by a priest and goddess. At the top of the shrine there is a lunar disk, a solar disk, and an eight-pointed star. These represent the gods Sin, Shamash, and Ishtar, respectively. The foundation of the divine dwelling is set on the heavenly sea, which is represented by wavy lines (bracket). The firmament (arrow and shaded) holds up this body of water. The four circular structures just above the firmament are the wandering stars (planets) Mercury, Mars, Saturn, and Jupiter. Notably, the Mesopotamian (Akkadian) word for "heaven" is *šamê*, and the individual words *ša* and *mê* mean "of water." Some Mesopotamian texts refer to rain being released

HISTORICAL CRITICISM

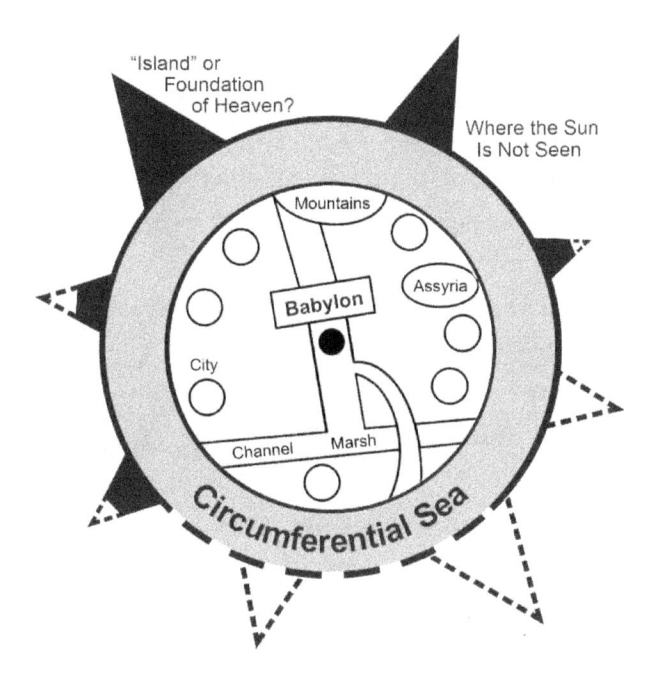

Figure 14-6. Ancient Mesopotamian Geography

through ducts in the firmament, and these are termed the "teats of heaven" and "breasts of heaven."[14]

Figure 14-6 is a drawing of the *Babylonian Map of the World*, which is dated about the sixth century BC.[15] It is important to emphasize that this is the entire earth according to these ancient Mesopotamians. From their ancient perspective, the city of Babylon was assumed to be at the centre of a circular flat earth surrounded by a circumferential sea (shaded). The two parallel lines going through Babylon represent the banks of the Euphrates River. Debate exists regarding the five (possibly eight?) triangular areas on this map. One suggestion is that they represent islands out at sea.[16] I suspect that these islands may also be pillars that support the solid dome of heaven (firmament) since they appear at the horizon.[17] And like the Egyptians, the Mesopotamians believed in an underworld and that the earth was set on water. In the *Enuma Elish*, written between the fourteenth to twelfth centuries BC, the creator god Marduk is identified as "lord of all the gods of heaven and underworld."[18] The beginning

105

of the *Bilingual Creation of the World by Marduk* describes the creation of the earth. Originating from about the third millennium BC, it records, "All the lands were sea . . . Marduk constructed a raft on the waters; he created dirt and piled it on the raft." [19] Marduk's placement of dirt on a floating raft indicates that the earth is flat and surrounded by water and that water is below the earth.

Historical criticism reveals that the ancient Egyptians and ancient Mesopotamians accepted a 3-tier universe. The question naturally arises: did the Holy Spirit-inspired writers of the Bible also accept this ancient conception of the structure and operation of the world? As we noted above, Scripture has an ancient understanding of reproductive biology similar to that of the Greek author Aeschylus. It is only reasonable that the Word of God includes the geography and astronomy-of-the-day in the ancient Near East.

Biblical Creation Accounts

In the light of historical criticism, we can ask some illuminating questions about the passages in Scripture that deal with the creation of the universe and living organisms. To use George Eldon Ladd's aphorism, are the accounts of origins in Genesis 1 and 2 "the Word of God given in the words of men in history?" Or to apply John Walton's saying, were the biblical creation accounts "written *for* us," but "not written *to* us?" And to be more precise, are these opening chapters in Genesis historically conditioned with an ancient science of origins?

In my opinion, the best way to answer these questions is to begin by examining statements in the Bible that refer to the structure and operation of the earth and heaven. Now I want you to be fully aware that I have an agenda in Hermeneutical Principles 15 and 16 as we explore whether or not a 3-tier universe appears in Scripture. There are three significant implications if the Word of God has this ancient geography and ancient astronomy.

First, the presence of a 3-tier universe in the Bible would be definitive proof that scientific concordism fails. In other words, the Holy Spirit's authorial intention was not to reveal modern scientific facts in Scripture prior to their discovery by modern science. If this is the case, then God accommodated in the process of biblical revelation and used the geography and astronomy-of-the-day in the ancient Near East.

Second, an ancient understanding of the structure and operation of the earth and heaven in Scripture raises challenging questions about God's creative action. Should the Bible present the Creator making a 3-tier universe *de novo* (quick and complete), it becomes clear that the Word of God does not reveal how he actually created the world, because we do not live in such a world. Consequently, the Lord's creative action in Scripture would have been filtered or accommodated through ancient scientific categories of origins, like *de novo* creation.

Finally, if the Bible features an ancient geography and an ancient astronomy, then it is only consistent and reasonable to suggest that it also has an ancient biology. Should this be the case, it would mean that the Word of God includes an ancient view of the origin of plants and animals . . . as well as humans. To be sure, proposing that the creation of men and women in Scripture reflects an ancient biological understanding of origins can be quite threatening to most Christians. But this is an idea that we need to consider seriously.

HERMENEUTICAL PRINCIPLE 15

The 3-Tier Universe: Ancient Geography

As we discovered in the previous hermeneutical principle, the ancient Egyptians and ancient Mesopotamians believed that the earth is flat, there is a solid firmament overhead holding up a heavenly sea, and an underworld exists below the surface of the earth. In Hermeneutical Principles 15 and 16, I will attempt to demonstrate that the Bible also has an ancient phenomenological perspective of the structure and operation of the earth and heaven. Stated more specifically, I will propose that the Word of God features a 3-tier universe with ancient geography and ancient astronomy as presented in Figure 15-1.[1] In proceeding through these two principles, it will be necessary to suspend our twenty-first century scientific ideas and read Scripture through an ancient mindset and picture the physical world through ancient eyes.

Let's begin by considering statements about the earth in the Bible. Taking into account their ancient phenomenological perspective, I will suggest that the biblical authors and their readers assumed that the earth: (1) is flat, (2) immovable, (3) has solid foundations that are set in water, (4) is surrounded by a circumferential sea, (5) has a circular outer boundary, (6) comes to an end at the shoreline of a circumferential sea, and (7) has an underworld below its surface. As we will see, these ancient geographical features make perfect sense from the limited knowledge and experience of ancient peoples. But more importantly, this ancient geography in Scripture is a vessel that delivers inerrant messages of faith.

THE 3-TIER UNIVERSE: ANCIENT GEOGRAPHY

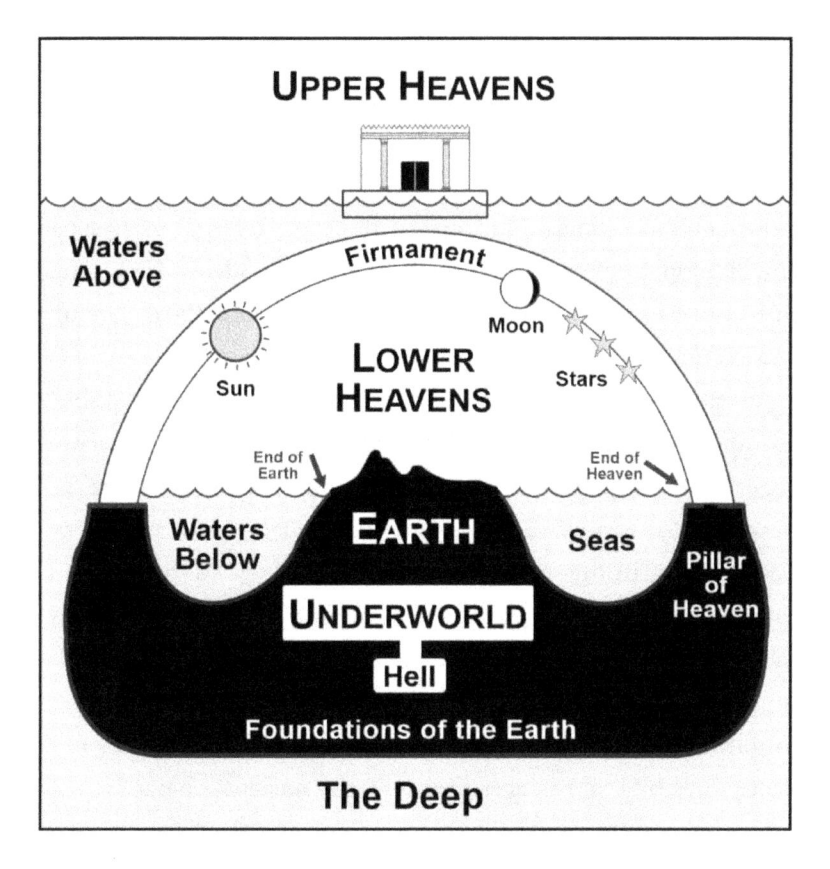

Figure 15-1. The Biblical 3-Tier Universe

1. The Flat Earth

I believe that one of the best arguments against scientific concordism is the structure and shape of the earth as described in the Bible. If it were the Holy Spirit's authorial intention to reveal scientific facts in Scripture well before their discovery by modern science, then I think it is reasonable to assume that he would have disclosed the shape of our home, planet earth.* In order to make this modern scientific idea understandable for ancient people, God could have compared the earth to something spherical like a ball or an apple.

---

* Some Christians might be quick to argue that Job 26:10 and Isaiah 40:22 reveal the earth is a spherical planet suspended in outer space. As we proceed, I will deal directly with these biblical verses.

109

However, the biblical evidence against a divine revelation about the earth being a sphere is overwhelming. In the Old Testament, the Hebrew word translated as "earth" in English Bibles is *'ereṣ*, and it appears about 2500 times. In the New Testament, the Greek term is *gē*, and it is found roughly 250 times. Not once is the earth referred to as a ball or sphere in Scripture. For me, this dearth (lack) of evidence for a spherical earth is a powerful argument against scientific concordism. I term this the "earth-dearth argument against concordism."

It is necessary to point out that there is no passage in the Bible that states *explicitly* the earth is flat. But as we will see with many scriptural passages in this hermeneutical principle, they only make sense in the context of a flat earth. This was an implicit geographical concept of the biblical authors. In the same way that we use the word "earth" and automatically assume a spherical planet, ancient people presupposed the earth was flat. This ancient scientific belief was quite reasonable. Anyone looking out from an elevated position like a mountaintop perceives the earth to be a level plain bordered by a flat circumferential horizon. Let's consider three biblical passages that *imply* this ancient geography.

Matthew 4:1-11 is a well-known passage that describes the temptation of Jesus by the devil. Verses 8 and 9 record, "Again, the devil took him [Jesus] to a very high mountain and showed him *all* the kingdoms of the world and their splendor. 'All this I will give you,' he [the devil] said, 'if you will bow down and worship me'" (my italics). The Greek word translated as "world" is *kosmos*, which in this passage means every place on earth where people live. But as everyone knows, there were great civilizations in China and Central America during the days of Jesus, and no matter how tall this mountain might have been, it was not possible for the Lord to see "*all* the kingdoms of the world." This passage therefore implies a flat earth in a 3-tier universe.

In Job 38:12-14, God challenges Job, "Have you ever given orders to the morning, or shown the dawn its place that it might take the earth by the edges and shake the wicked out of it? The earth takes shape like clay under a seal." The expression "take the earth by the edges" clearly implies a flat earth with ends, because there are no edges on a spherical earth. Picture someone grabbing the edges of a towel and shaking it. Ancient geography is also reflected in the sentence, "The earth takes shape like clay under a seal." In the ancient world, most seals were flattened impressions made of clay or wax. This passage only makes sense if the earth is assumed to be flat.

The Book of Revelation deals with the end of time and the Second Coming of Jesus. Revelation 1:7 states, "Look, he [Jesus] is coming with the clouds, and '*every* eye will see him, even those who pierced him;' and *all* the peoples of the earth 'will mourn because of him'" (my italics; cf. Acts 1:11). In the context of a spherical earth, it is not possible for "all the peoples on earth" to see Jesus, because one side of the earth would be hidden from him. It is only from an ancient phenomenological perspective in a 3-tier universe with a flat earth that "every eye" could see the Lord in this passage.

If we read Matthew 4:1-11, Job 38:12-14, and Revelation 1:7 in the light of the Message-Incident Principle, we can avoid an unnecessary conflict between the Bible and modern geography. In Matthew 4, the ancient geography is a vessel that delivers the inerrant message, "Worship the Lord your God, and serve him only" (v. 10). The spiritual truth in Job 38 is that God is the Creator, not Job. And Revelation 1 assures us that Jesus will come again at the end of the world and everyone will know it.

2. The Immovability of the Earth

Ancient people believed that the earth was stationary and did not move. From their narrow scope of cognitive competence, this was a very logical idea. Even today we do not sense that the earth is spinning on its axis at 1000 miles per hour and traveling around the sun at 70,000 mph. Belief in the immobility of the earth lasted as late as the seventeenth century. This notion was challenged by Galileo at that time, and it was later rejected when the motion of the earth was established as a scientific fact, after the construction of powerful telescopes in the nineteenth century.

Three biblical verses that state the earth is immovable are 1 Chronicles 16:30, Psalm 93:1, and Psalm 96:10. In the original Hebrew language, they repeat word-for-word, "The world is firmly established, it cannot be moved" (NIV 1978). The Hebrew noun *tēbēl* translated as "world" in these verses can refer simply to the earth or to both heaven and earth (e.g., Ps. 89:11). Throughout their lifetime, ancient Hebrews perceived that plains and mountains remained in place, and they reasonably assumed that the earth was stationary.

The incidental ancient geography in these three verses delivers the message of faith that the God of the Bible is the Creator and Judge of the world. For example, Psalm 96:10 states, "Say among the nations, 'The Lord reigns.' The world is firmly established, it cannot be moved; he will judge the peoples with equity." The Old Testament often presents physical reality and moral reality together, placing these two concepts parallel to each other. This verse reveals that the immovability of God's creation is an assurance of the stability of God's justice.

3. The Foundations of the Earth Set in Water

Ancient individuals used familiar objects to explain their experience that the world was stable and did not move. In the Bible, the authors often employed the engineering term "foundation/s" of a building (1 Kgs. 7:10; Ezr. 5:6) to conceptualize the immovability of the earth. For example, Psalm 104:5 states, "He [God] set the earth on its foundations; it can never be moved." Micah 6:2 also refers to the "everlasting foundations of the earth." And in Job 38:4 and 6, the Creator challenges Job. "Where were you when I laid the earth's foundation? . . . On what were its bases sunk, or who laid its cornerstone?" (NRSV)

To answer the question "on what were its [the earth's] bases sunk?" it must be remembered that according to ancient Near Eastern science, the earth was placed on water. We saw this ancient geographical notion in Principle 14 with the Mesopotamian account the *Bilingual Creation of the World by Marduk*. In creating the earth, "Marduk constructed a raft on the waters; he created dirt and piled it on the raft." [2] The ancient Egyptians also believed that the earth was set upon a deep body of water that they called "Nun." A hymn praises their creator god Ptah, "who founded the earth . . . who surrounded it with Nun, and the sea."[3]

The ancient geographical idea that the bases or foundations of the earth were set on water appears in the Bible. Psalm 24:2 asserts, "The earth is the Lord's, and everything in it, the world, and all who live in it; for he has founded it on the seas and established it upon the waters." Psalm 136:6 also states that God "spread out the earth upon the waters." And the Second Commandment in Exodus 20:4 orders, "You shall not make for yourself an idol, whether in the form of anything that is in heaven above, or that is on the earth beneath, or that is in *the water under the earth*" (NRSV, my italics).

This body of water below the earth is often referred to in Scripture as "the deep" (Hebrew *tehōm*). Ancient people were aware that fresh water came out of the depths of the earth, and the Bible calls these the "fountains of the deep" and the "springs of the great deep." For example, Proverbs 8:28 states that God "fixed securely the fountains of the deep" when he created the world. During the first day of the biblical flood, Genesis 7:11 records that "all the springs of the great deep burst forth" and water began to cover the earth (also Gen. 8:2). Figure 15-1 identifies the waters of the deep below the foundations of the earth.

But more importantly, the incidental ancient geography in these biblical passages above delivers inerrant spiritual truths. The verses referring to foundations of the earth (Ps.104:5; Job 38:4, 6) and those with the earth set upon water (Ps. 24:2; Ps. 136:6) both affirm that God is the Creator and Sustainer of the earth. And Exodus 20:4 warns us not to make idols of anything found in the entire universe, because there is only one God, the God of the Bible.

4. The Circumferential Sea

Ancient Near Eastern people believed that the earth was surrounded by a circular sea. Though such an idea makes little sense to us today, this ancient geographical concept was perfectly reasonable considering

THE BIBLE & ANCIENT SCIENCE

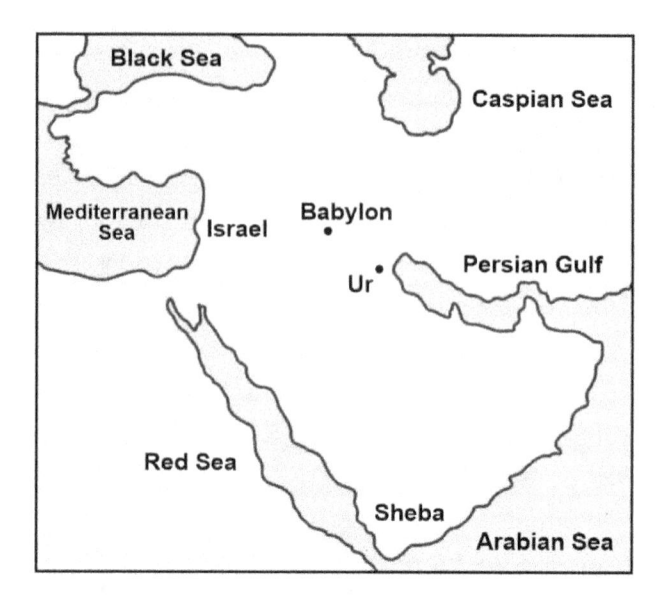

Figure 15-2. Geography of Ancient Near East

two ancient phenomenological experiences. First, the visual effect of the horizon gives the impression that the entire world is enclosed within a circular boundary. Second, as Figure 15-2 reveals, traveling in almost any direction in the ancient Near East brought a person to the shoreline of a large body of water. Historical criticism confirms that a circumferential sea surrounding a circular earth was part of the geography-of-the-day about 2500 years ago in ancient Egypt (Fig. 14-4, p. 103) and ancient Mesopotamia (Fig. 14-6, p. 105).

As we can see in Figure 15-1, it is necessary to distinguish the "waters below" the firmament (which became the circumferential sea) from the heavenly "waters above" the firmament and the waters of the deep underneath the earth. In this way, we can have a better understanding of the formation of the earthly sea as described on the third day of creation in Genesis 1:9-10. "And God said, 'Let the water under the sky [i.e., the firmament] be gathered to one place, and let dry ground appear.' And it was so. God called the dry ground 'land,' and the gathered waters he called 'seas.'"

Two biblical creation accounts outside of Genesis 1 shed light on the origin of the circumferential sea. We noted earlier, the opening scene in Genesis 1 is Genesis 1:2. This verse depicts the earth entirely covered by water and the deep, as shown in Figure 12-4 (p. 88). The Proverbs 8:22-31 account of creation records in verses 27 and 28 that God "inscribed a circle on the face of the deep, when he made *firm* the skies above" (NASB; my italics). In other words, the Creator drew a circle on the waters mentioned in Genesis 1:2 and outlined the rim of the firmament at the boundary of the horizon where the dome of heaven and the sea meet.

Similarly, in the Job 26:7-14 creation account, God also defines the horizon at the outside edge of the earthly sea. Verse 10 states that the Creator "inscribed a circle on the surface of the waters, at the boundary of light and darkness" (NASB). This outer border of the sea is where ancient people perceived the sun to rise and set at the base of the firmament. The Hebrew noun translated as "circle" in both Proverbs 8:27 and Job 26:10 is *ḥûg*, and it refers to a flat two-dimensional geometric figure. In some English Bibles, this word is rendered as "horizon" (NIV) or "compass" (KJV) to indicate its association with a flat surface.

Proverbs 8:27-28 and Job 26:10 offer insights to understand the origin of the sea on the third day of creation. The gathering of the waters "to one place" under the dome of the sky/firmament refers to the formation of a single body of water—the circumferential sea at the border of the horizon. And the appearance of dry land implies that the earth is surrounded by water. A way to picture this creative scene is that God makes a circular trench in the formless earth of Genesis 1:2 for the sea to collect near the horizon, and then he raises dry land in the middle of these "waters below" the sky/firmament.

Of course, there is a central spiritual truth in these biblical passages that deal with the sea. Even though the Holy Spirit accommodated and descended to the level of the biblical writers and allowed them to use the ancient understanding that a circumferential sea surrounded the earth, the inerrant message of faith grasped by Christians throughout history is

THE BIBLE & ANCIENT SCIENCE

that the Lord made the earthly sea. This is the power of the divine revelation in the Word of God. The spiritual truths in Scripture are eternal, and they transcend time and place and speak to every generation of believers.

5. The Circle of the Earth

People in the ancient Near East believed in a circumferential sea, and it was only logical that its inner border surrounded a flat earth that was circular. As we saw previously in the ancient Mesopotamian and ancient Egyptian maps of the world (Figs. 14-4 and 14-6, pp. 102 and 105), this was the geography-of-the-day. And as we noted, one inscription from Egypt in ancient times even refers to the "circle of the earth."[4]

The term "circle of the earth" is explicitly stated in Scripture. Isaiah 40:22 records, "He [God] sits enthroned above the circle of the earth." Of course, just about every Bible-believing Christian knows this verse, and many use it in an attempt to prove that God revealed a scientific fact in Scripture hundreds of years before it was discovered by science. They contend that the "circle" in this verse refers to the outline of planet earth from the perspective of outer space. You can now appreciate that such an interpretive approach is rooted in scientific concordism.

The use of Isaiah 40:22 in this way is a classic example of a proof-text interpretation. It tears a part of this verse out of its ancient context, and then manipulates it to mean something that was never intended by the Divine Author or the human author. Here is the *entire* verse: "He [God] sits enthroned above the circle of the earth, and its people are like grasshoppers. He stretches out the heavens like a canopy, and spreads them out like a tent to live in" (cf. Ps 33:13). In other words, the structure of the universe is like a tent with a domed canopy overhead and a flat floor below. This verse is clearly consistent with a 3-tier world.*

In addition, the word translated as "circle" in Isaiah 40:22 is the Hebrew noun *ḥûg*, and as we have seen, it refers to a flat two-

---

*Another well-known concordist proof-text is Job 26:7. See Appendix 3: Do Isaiah 40:22 & Job 26:7 Refer to a Spherical Earth?

dimensional geometric figure, not a three-dimensional sphere. If God through the prophet Isaiah had intended to reveal that the earth was spherical, he could have used the Hebrew word *dur*, referring to a ball, as he did in Isaiah 22:18. "He [God] will roll you [Jerusalem] up tightly like a ball and throw you into a large country." But the Holy Spirit did not inspire Isaiah to do so in order to disclose a modern scientific fact.

The circular earth is also implied elsewhere in Scripture. In Daniel 4:10-11, the King of Babylon describes a dream that compared him to a tree. "I looked, and there before me stood a tree in the middle of the land. Its height was enormous. The tree grew large and strong and its top touched the sky [i.e., the firmament]; it was visible to the ends of the earth." This passage makes sense, considering the Babylonian world map in Figure 14-6 (p. 105). It is only in the context of a circular flat earth that "the ends of the earth" could be seen from this tree. If the earth was spherical, this would not be possible, because one side of the earth would always be hidden no matter how tall the tree might be.

But more importantly, Isaiah 40:22 and Daniel 4:10-11 reveal inerrant spiritual truths. The prophet Isaiah underlines that God is the Creator of a world that he made specifically for us to live in. And the Lord graciously warned the King of Babylon in a dream that if he did not repent of his sins (v. 27), he would be cut down like a tree.

6. The Ends of the Earth

As we noted above, people traveling in any direction in the ancient Near East were likely to come to the end of dry land and to the shore of what they perceived to be a circumferential sea. It was quite reasonable for them to refer to this shoreline as the "ends of the earth." This ancient geographical term appears nearly fifty times in the Bible.

For instance, in Matthew 12:42 Jesus proclaims, "The Queen of the South will rise at the judgment with this generation and condemn it; for she came from the ends of the earth to listen to Solomon's wisdom, and now something greater than Solomon is here." The Lord was referring to the Queen of Sheba, and her land was in the southwest corner of the

Arabian Peninsula, bordered by the Arabian and Red seas. From an ancient geographical perspective, this region was at the ends of the earth as shown in Figure 15-2.

Similarly, Isaiah 41:8-9 states, "But you, Israel, my servant, Jacob, whom I have chosen, you descendants of Abraham my friend, I took you from the ends of the earth, from its farthest corners I called you." We know from Genesis 11:31 that Abraham came from the city of Ur, which was at the shore of the Persian Gulf. Again from an ancient phenomenological perspective, it made sense to say that Abraham once lived at the "ends of the earth," because Ur was at the end or edge of their world as identified in Figure 15-2.

Notably, Isaiah 41:9 refers to the "farthest corners" of the earth, possibly creating a conflict with the ancient notion that the earth was circular. In addition, there are four other verses in Scripture that refer to the "four corners of the earth/land." For example, Revelation 7:1 states, "After this I saw four angels standing at the four corners of the earth, holding back the four winds of the earth" (cf., Rev. 20:8; Isa. 11:12; Ezek. 7:2). However, the Hebrew noun *kānāp* in the Old Testament and the Greek *gōnia* in the New Testament can also mean "edge" or "extremity." Therefore, Isaiah 41:9 is best translated, "I took you from the ends of the earth, from its extreme edges I called you."

In addition, Revelation 7:1 refers to "four winds." The categorization of the wind (and the earth) into north, south, east, and west is a phenomenological perception and reflects human sensory experience. We see in front of us, we have an ear on our left and right, and we have no visual perception behind us. Thus, there is no conflict in biblical passages that refer to the earth being circular and having four extremities. This reflects the interaction of two phenomenological factors—the circumference of the horizon and the anatomical position of human senses.

Yet once again, the purpose of the Bible is to reveal messages of faith. In Matthew 12:42, Jesus is stating that his wisdom is greater than that of Solomon, and that there will be a Final Judgment. And Isaiah 41:8-9 offers the encouraging spiritual truth that God personally called Abraham and Israel in order that "all peoples on earth will be blessed through you" (Gen. 12:3). And indeed, we enjoy this blessing as Christians!

7. The Underworld

To the surprise of many Christians, the Bible refers to a region under the surface of the earth known as the "underworld." It is the lower level of the 3-tier universe. Two ancient phenomenological experiences led to this ancient geographical idea. First, the sun sets in the west and then rises in the east. It was perfectly reasonable to assume that the sun travelled underneath the earth at night. Second, ancient people experienced earthquakes and it was logical to believe that there were powerful gods or demons in the underworld and that they could shake the earth, like divine beings thundering in heaven and launching lightning. Notably, the Middle East is susceptible to earthquakes because three major continental (tectonic) plates—African, Eurasian, and Arabian—contact each other and their movement often produces earthquakes.

According to ancient Near Eastern people and the inspired biblical writers and their readers, the underworld is where the souls (spirits) of men and women went after their death. In the Old Testament, the Hebrew noun *she'ōl* appears sixty-five times and refers to the underworld. For example, in a judgment against the King of Babylon, Isaiah 14:9 declares, "The realm of the dead below [*she'ōl*] is all astir to meet you at your coming; it rouses the spirits of the departed to greet you." Similarly, God states in Deuteronomy 32:22, "For a fire will be kindled by my wrath, one that burns down to the realm of the dead below [*she'ōl*]."

The context in about half of the occurrences of *she'ōl* indicates that it is subterranean. In Amos 9:2, God proclaims, "Though they dig down to the depths below [*she'ōl*], from there my hand will take them. Though they climb up to the heavens above, from there I will bring them down." The expression to "dig down" definitely refers to tunneling downward through the earth. This verse fits perfectly within the context of a 3-tier universe with an underworld below the surface of the earth and a heavenly realm above. In a similar context, Psalm 139:7-8 states, "Where can

THE BIBLE & ANCIENT SCIENCE

I go from your [God's] Spirit? Where can I flee from your presence? If I go up to the heavens, you are there; if I make my bed in the depths [*she'ōl*] you are there."

The Old Testament seems to distinguish a lower region in the underworld termed "Abaddon" and sometimes refers to the "depths of the Pit." The Hebrew noun *'ăbaddōn* means "place of destruction." Psalm 88:3-6 and 11 state, "For my soul is full of troubles, and my life draws near to Sheol. I am counted among those who go down to the Pit; . . . like those forsaken among the dead . . . You [God] have put me in the depths of the Pit, in regions dark and deep . . . Is your steadfast love declared in the grave, or your faithfulness in Abaddon?" (NRSV). This distinction of two regions in the underworld seems to appear in Proverbs 27:20, "Sheol and Abaddon are never satisfied," and Proverbs 15:11, "Sheol and Abaddon lie open before the Lord" (NRSV).

The New Testament mentions the underworld about twenty times and uses different terms and expressions in referring to this region. As noted previously, a more accurate translation of Philippians 2:9-10 renders the Greek *katachthoniōn* in verse 10 as: "Therefore, God exalted Jesus to the highest place and gave him the name that is above every name, that at the name of Jesus every knee should bow, [1] in heaven and [2] on earth and [3] down in the underworld [*katachthoniōn*]."

The Greek expression *hupokatō tēs gēs* also refers to the subterranean world. The adverb *hupokatō* means "below," "under," and "underneath." The word *tēs* is the definite article "the" and *gēs* is a form of *gē*, the term for "earth." In Revelation 5:13 the apostle John writes, "Then I heard every creature [1] in heaven and [2] on earth and [3] under the earth [*hupokatō tēs gēs*] and on the sea, and all that is in them, saying, 'To him [Jesus] who sits on the throne and to the Lamb be praise and honor and glory and power, for ever and ever!'" Like Philippians 2:9-10, this verse praises the Lord and makes perfect sense in a 3-tier universe.

The Greek noun *hādēs* is another term for the underworld. In Matthew 11:23, Jesus declares, "And you, Capernaum, will you be lifted up to the heavens? No, you will go down to Hades." This verse is quite logical within the context of a world with three tiers. And in one of the most important passages in the New Testament, the apostle Peter states to Jesus in Matthew 16:16, "You are the Messiah, the Son of the living God." The Lord then replies in verse 18 that "upon this rock I will build my church, and the gates of Hades will not overcome it." In other words, Christianity is founded on the bedrock inerrant spiritual truth that Jesus is the son of God and our Savior. The Lord has overcome the power of death because the gates of the realm of the dead cannot hold back those who believe in Jesus.

It is important to point out that the Lord identifies the location of the underworld in Matthew 12:40. "For as Jonah was three days and three nights in the belly of a huge fish, so the Son of Man [Jesus] will be three days and three nights in the *heart of the earth*" (my italics). Of course, Jesus is referring to his death and the three days before his bodily resurrection. In accommodating to the ancient geography of his listeners, the Lord described the underworld as being in the middle of the earth.

The New Testament also recognizes an area in the depths of the underworld reserved for the souls of evil humans and demonic spirits. Often translated as "hell" in English Bibles, three Greek terms are used to designate this region—*tartarus*, *abussos*, and *gehenna*. 2 Peter 2:4 states that "God did not spare the angels when they sinned, but sent them into hell [*tartarus*], putting them in chains of darkness [or gloomy dungeons]." In Revelation 20:1-3, the apostle John has a vision of the end times. "And I saw an angel coming down out of heaven, having a key to the Abyss [*abussos*] and holding in his hand a great chain. He seized the dragon, that ancient serpent, who is the devil, or Satan, and bound him a thousand years. He threw him into the Abyss and locked and sealed it over him." The connection between the Old Testament's Abaddon and the New Testament's Abyss appears in Revelation 9:11 with the phrase, "The angel of the Abyss, whose name in Hebrew is Abaddon." Figure 15-1 identifies the lower region of hell in the underworld.

It must be emphasized that Jesus definitely believed in the existence of hell. Of the twelve times *gehenna* appears in the New Testament, eleven are from teachings of the Lord. In Luke 12:5 Jesus cautions, "Fear him who, after your body has been killed, has the authority to throw you into hell [*gehenna*]." The Lord also recognized that the underworld had two distinct regions in his parable of Lazarus and the rich man, in which he describes their fate after death (Lk. 16:19-31). There was "a great chasm" (v. 26) between an upper region with Lazarus at Abraham's side and a lower region of fire and torment with the rich man. As well, Jesus acknowledged the existence of the Abyss. In Luke 8:26-39, he heals the demon-possessed man and agrees not to send the demons "into the Abyss [*abussos*]" (v. 31), but instead into a herd of pigs. And 1 Peter 3:18-20 suggests that the Lord after his death preached to the souls of those in hell. "He [Jesus] was put to death in the body but made alive in the Spirit. After being made alive, he went and made proclamation to the imprisoned spirits—to those who were disobedient long ago."

Though the notion of an underworld in the Bible is not commonly known among Christians today, this ancient geographical notion is used to reveal inerrant spiritual truths. Amos 9:2 and Psalm 139:7-8 emphasize that there is no place in the world that we can hide from God. Philippians 2:10-11 and Revelation 5:13 disclose that Jesus is Lord over everyone in the universe, including the spirits/souls of the dead. And most importantly, in Luke 8:31, 12:5 and 16:26, Jesus reveals that there is life after death, and that hell really does exist, even though we may not know its actual location. But the Good News of the Gospel is that God offers us the hope of eternal life, because Matthew 16:18 states the "gates of Hades" will not hold us back from being with our Lord and Savior Jesus Christ. Amen!

HERMENEUTICAL PRINCIPLE 16

The 3-Tier Universe: Ancient Astronomy

The inspired authors of Scripture and their readers looked up at the heavens and understood its structure and operation from an ancient phenomenological perspective. This led them to assume: (1) the earth and circumferential sea are enclosed by a solid dome-like structure known as the "firmament," (2) the heavens (firmament) have ends and are set on pillars/foundations, (3) a heavenly sea is held up by the firmament, (4) the sun, moon, and stars are placed in the firmament, (5) the sun actually moves across the sky every day, (6) at the end of time stars will fall to the earth and the heavens will be rolled up, and (7) the heavens are made up of two physical regions: the lower heavens contain the atmosphere and the firmament, and the upper heavens is the realm of God and the angels and include his divine dwelling. These ancient scientific concepts are components of the 3-tier universe as shown in Figure 15-1 on page 109.

Similar to the ancient geography in the Word of God, these ancient astronomical features made perfect sense from the narrow scope of cognitive competence of the biblical writers and their generation. Had we lived during that time, we too would have accepted this ancient science. This is another example of God accommodating to the level of ancient people in the process of biblical revelation. But more importantly, this ancient understanding of the heavens in Scripture is ultimately incidental and delivers life-changing, inerrant spiritual truths.

1. The Firmament

In my opinion, the best evidence for ancient science in the Bible is the firmament. Regrettably, most Christians today have no idea what this term actually means. From an ancient phenomenological perspective, the vault of the sky and the circumference of the horizon give the impression there is a firm immovable structure overhead, like an inverted bowl.* Moreover, the blue of the sky led to the logical idea that there was a heavenly sea of water held up by a transparent firmament.

The firmament was created on the second day of creation in Genesis 1:6-8. "And God said, 'Let there be a firmament between the waters to separate water from water.' So God made the firmament and separated the water under the firmament from the water above the firmament. And it was so. God called the firmament 'heaven.'"[1] The firmament also appears in the first verse of the beloved Psalm 19. "The heavens declare the glory of God; the firmament proclaims the work of his hands."

The word "firmament" is a translation of the Hebrew noun *rāqîaʿ*. It is related to the verb *rāqaʿ* which means "to flatten," "hammer down," and "spread out." In particular, this verb has the sense of flattening something solid, and it is found in the context of pounding metals into thin plates. For example, Exodus 39:3 states, "They hammered out thin sheets of gold." The related noun *riqqûaʿ* refers to metal sheets. As Numbers 16:38 commands, "Hammer the [metal] censers into sheets to overlay the altar." Notably, the verb *rāqaʿ* is even found in a passage that refers to the creation of the sky, which is understood to be solid and similar to a metal. Job 37:18 asks, "Can you join him [God] in spreading out [*rāqaʿ*] the skies, hard as a mirror of cast bronze?"

It must be noted that the word translated as "skies" in Job 37:18 is not *rāqîaʿ*, but the noun *sheḥāqîm*. This is another Hebrew term for the

---

* I suspect several of you have noticed that the word "firmament" has two distinct meanings. In older texts like the Bible and the works of the ancient Egyptians and Mesopotamians, it refers to a hard dome in heaven (Figs. 14-3, 14-5, 15-1). In later literature such as the writings of Luther, Copernicus, and Galileo, the firmament is understood to be the solid sphere of heaven, as seen in geocentric and heliocentric theories of the universe (cover of this book, Figs. 9-1, 9-2, 9-3).

firmament. It also appears in Job 37:21, "Now no one can look at the sun, bright as it is *in* the skies [*shehāqîm*], after the wind has swept them clean [of clouds]" (my italics). This is consistent with the fourth day of creation when God placed the sun *in* the *rāqîa'*. The singular form of *shehāqîm* is *shahaq* and it derives from the verb *shāhaq* that has meanings "to beat fine" and "pulverize to dust." It may be that the countless speckles of stars in the night sky, and especially the "dusty" appearance of the Milky Way, led ancient people from their ancient phenomenological perspective to believe that the firmament was dusted with fine lights.

Early translations of the Bible also indicate that the Hebrew noun *rāqîa'* refers to a solid structure over the earth. The Septuagint is a Greek translation of the Old Testament dated around 250 BC. It translates *rāqîa'* as the noun *stereōma*. We noted in Hermeneutical Principle 13 that the adjective *stereos* means "hard," "firm," and "solid." Therefore, *stereōma* signifies that the firmament is the hard and solid part of the sky. The Vulgate is a Latin Bible that served the church for nearly one thousand years. Translated around AD 400, this version renders *rāqîa'* as *firmamentum*. This noun is related to the adjective *firmus* which means "hard," "firm," and "solid." Finally, the English King James Version was published in 1611. It translates *rāqîa'* into the term "firmament."

Historical criticism further assists to determine the meaning of the Hebrew word *rāqîa'*. As we noted earlier, ancient Near Eastern images reflect this ancient astronomy. An Egyptian depiction of the universe between 1570 and 1085 BC presents a solid domed firmament speckled with stars above the earth (Fig. 14-3, p. 102). The Mesopotamian understanding of the structure of heaven found in the Shamash Tablet, dated around 850 BC, also has a solid firmament. It is connected to wandering stars (planets) and holds up a heavenly sea overhead (Fig. 14-5, p. 104).

To summarize, the original meaning of *rāqîa* within its biblical context, the related words found in Scripture, the early biblical translations of this term, and historical criticism all indicate that this Hebrew noun refers to a firm immovable structure above the earth. In recent years, modern translations of the Bible have reflected this ancient astronomical notion. The New Revised Standard Version (1991) translates *rāqîa* as "dome" and the New International Version (2011) employs "vault." Personally, I prefer the traditional term "firmament" because it fully captures the original meaning of a firm solid structure above the earth.

Of course, the ultimate purpose of the Bible is to reveal messages of faith. On the second day of creation in Genesis 1, the ancient notion of a firmament is an incidental vessel that delivers the spiritual truth that God created the perceived dome of heaven. Similarly, Psalm 19:1 declares that this structure "proclaims the work of his hands." These inerrant divine revelations remain true for us today. The appearance of the "dome" of the sky was made by our Creator, and it continues to be a natural revelation reflecting the intelligent design that God has inscribed in the heavens.

2. The Ends & Pillars/Foundations of the Heavens

Ancient Near Eastern people believed in a solid firmament above the earth. The visual impact of the horizon led them to the reasonable conclusion that the hard dome of the heavens had actual "ends" where the firmament met the outer boundary of the circumferential sea. From their narrow scope of cognitive competence, this heavenly dome appears to be solid and immovable. It was logical then for ancient individuals to conclude that the firmament was set on pillars or foundations like those upon which the earth was placed (Fig. 15-1, p. 109).

The Bible refers to the ends of the heavens. In teaching about the end of time, Jesus states in Matthew 24:31 that "he will send his angels with a loud trumpet call, and they will gather his elect from the four winds, from one end of the heavens to the other." From the perspective of a 3-tier universe, this gathering of the Lord's people would extend across the entire world enclosed under the firmament. Similarly, Psalm 19:6 asserts that the sun "rises at one end of the heavens and makes its circuit to the other." In other words, for ancient people the horizon where the sun rises and sets was understood to be the ends of the heavens.

Scripture also mentions the pillars or foundations of the heavens. Like the term the "foundations of the earth," it uses these engineering concepts to explain the immovability of the dome of the firmament. However, God could shake these solid bases. As Job 26:11 records, "The pillars of the heavens quake, aghast at God's rebuke." 2 Samuel 22:8 also asserts, "The earth trembled and quaked, the foundations of the heavens shook; they trembled because he [God] was angry." The biblical writers and their readers undoubtedly experienced earthquakes, and the idea that the heavens and the earth and their foundations were being shaken by God made perfect sense to them.

These passages referring to the ends and pillars/foundations of the heavens transport inerrant spiritual truths. In Job 26 and 2 Samuel 22, the message of faith is that the Creator is in total control of the world, and that he can shake it whenever he wants. Psalm 19 reveals that God created the sun and it declares his glory. And in Matthew 24:31, Jesus promises that at the end of time he will gather his people from all over the world.

3. The Heavenly Sea

The belief that there existed a large body of water over the earth was reasonable from an ancient phenomenological perspective. Rain falls from above and the color of the dome of the sky is a changing blue that resembles a lake or sea. As we noted in Genesis 1:7, God created a solid firmament on the second day of creation and "separated the water under the firmament from the water above the firmament." Consequently, these two bodies of water are sometimes called the earthly "waters below" and the heavenly "waters above" as identified in Figure 15-1 (p. 109).

Interestingly, the Hebrew word for water is *mayim* and it has a special ending not found in English. The dual ending *-ayim* is used for things in nature occurring in pairs. For example, *yādayim* refers to a pair of hands, *ragelayim* to both feet, and *kenāpayim* to the two wings of a bird. Thus, the noun *mayim* suggests that there are two major bodies of water in the universe—the sea above in heaven and the sea below on earth. In the Old Testament, *mayim* appears about 600 times, and it is always in the dual, never in the singular or the plural.

Historical criticism affirms that the notion of a sea in heaven was accepted in the ancient Near East. As we have seen in the diagram of the Egyptian 3-tier universe (Fig. 14-3, p. 102), the sun god Re appears in a boat two times. This clearly demonstrates that the ancient Egyptians believed in the existence of a heavenly sea. Similarly, in the Mesopotamian Shamash Tablet (Fig. 14-5, p. 104), the wavy lines between the firmament and the foundation of the divine dwelling also reveal that these ancient peoples accepted a sea of water above the earth in the heavens. And as we noted earlier, the word for "heaven" in Akkadian is *šamê* and the words *ša* and *mê* mean "of water." The Mesopotamians believed rain fell to earth from channels in the firmament, which they called the "teats of heaven" and "breasts of heaven."[2]

Equipped with this ancient astronomical concept of a heavenly sea, we can view some biblical passages in a new light. Psalm 104:2-3 states, "He [God] stretches out the heavens like a tent and lays the beams of his upper chambers on their waters." The tent metaphor in verse 2 points to a universe with a domed heaven and a flat earth. And like the Shamash Tablet with the foundation of the Shamash's shrine being set upon the heavenly sea (Fig. 14-5, p. 104), the "beams" of God's "upper chambers" in his heavenly dwelling are also placed on the "waters above" (Fig. 15-1, p. 109).

The sea in heaven also appears in Psalm 148:3-4. "Praise him [the Lord], sun and moon; praise him, all you shining stars. Praise him, you highest heavens and you waters above the skies." In other words, the psalmist believes that there exists a body of water above the dome of the skies (firmament). As a final example, Jeremiah 10:12-13 asserts that God "stretched out the heavens by his understanding. When he thunders, the waters in the heavens roar." This biblical author clearly pictures God thundering over the heavenly sea and causing it to roar.

Now I want you to be aware that some Christians argue that the water referred to in Genesis 1:7, Psalm 104:2, Psalm 148:4, and Jeremiah

10:13 is water vapor or clouds in the atmosphere. However, ancient Hebrew has specific words meaning vapor/mist (*'ēd*, Prov. 21:6) and cloud (*nāśî'*, Ps. 135:7; *'ānān*, Gen. 9:14). If the intention of the biblical authors was to refer to water vapor or clouds in these verses, then they would have used these terms. But they didn't. Instead, these inspired writers used *mayim*, the Hebrew word for liquid water. This is the very same term used in Genesis 1:10 where God called the "gathered waters [*mayim*]" the "seas."*

To conclude, the writers of Scripture and their readers believed the "waters above" is a literal and actual body of liquid water in the heavens. This is another instance of biblical accommodation. Today we know the blue of the sky to be a visual effect due to the scattering of shortwave light in the upper atmosphere, but ancient people had no way of knowing this scientific fact. Despite these differences in understanding the structure of the heavens, the heavenly sea found in Scripture delivers the same inerrant spiritual truth to all generations—the God of the Bible is the Creator of the huge blue phenomenon that we see above the earth.

4. The Sun, Moon & Stars in the Firmament

From an ancient phenomenological perspective, it appears as though the heavenly bodies are positioned in or near the firmament. Historical criticism affirms that this was the astronomy-of-the-day in ancient Egypt and ancient Mesopotamia as shown in Figures 14-3 (p. 102) and 14-5 (p. 104), respectively. This ancient scientific understanding of the sun, moon, and stars being placed in the firmament is mentioned three times on the fourth day of creation in Genesis 1:14-19.

> And God said, "Let there be lights in the firmament of the heaven to separate the day from the night, and let them serve as signs to mark sacred times, and days and years, and let them be lights in the firmament of the heaven to give light on the earth." And it was so. God made two great lights—the

---

* In Appendix 2: The "Waters Above" & Scientific Concordism, I respond in more detail to two popular interpretations of the "waters above" in Genesis 1.

greater light to govern the day and the lesser light to govern the night. He also made the stars. God set them in the firmament of the heaven to give light on earth, to govern the day and the night, and to separate light from darkness. And God saw it was good.

The purpose of this biblical passage is not to reveal astronomical facts. Rather it offers a radical spiritual message to the Hebrews and the ancient nations surrounding them. This is a blunt criticism of the pagan religions and their belief that the sun, moon, and stars were gods. Under the inspiration of the Holy Spirit, the writer of Genesis 1 strips these astronomical bodies of their so-called "divine" status and turns them into mere creations of the God of the Hebrews. In this way, the Bible desacralizes (or demythologizes) the natural world.

Even more radically, Genesis 1:14-19 turns these so-called "gods" into servants! Instead of men and women serving heavenly bodies as demanded by astral religions, the inspired author states that the sun, moon, and stars were made to serve humanity "to mark sacred times, and days and years." The Bible puts the heavenly bodies in their proper place. They have value because they are God's "good" creations, but they are definitely not gods worthy of worship. The eternal message of faith to everyone is this: never serve the sun, moon, and stars. Let them serve you!

5. The Daily Movement of the Sun across the Sky

It was a common belief up to the seventeenth century that the sun literally traveled across the sky from east-to-west every day. This was quite logical because people before this time viewed the world through a narrow scope of cognitive competence. Like us, in the morning they saw the sun rise out of the east, and then "set" in the west during the evening. As well, they never felt the earth rotating on its axis. The Bible refers to "sunrise" and "sunset" over sixty times. In Hermeneutical Principle 5, we examined some of these passages and it is worth revisiting them in light of the 3-tier universe shown in Figure 15-1 (p. 109).

Ecclesiastes 1:5 records, "The sun rises and the sun sets, and hurries back to where it rises." Psalm 19:4-6 states, "In the heavens God has pitched a tent for the sun. It is like a bridegroom coming out of his chamber, like a champion rejoicing to run his course. It rises at one end of the heavens and makes its circuit to the other." And Psalm 113:3 proclaims, "From the rising of the sun to the place where it sets, the name of the Lord is to be praised."

In these three passages, the sun appears at one end of the horizon next to the base of the firmament, then moves across the dome of the firmament, and finally disappears below the horizon at the other end of the firmament. It then travels below the earth to return to the place where it rises again. From an ancient phenomenological perspective, belief in the literal and actual movement of the sun every day across the sky is very reasonable. As we have noted, this ancient astronomical idea was so persuasive that it lasted as late as the seventeenth century when it was challenged by Galileo.

Similarly, in Matthew 5:45 Jesus taught his disciples that our Father in heaven "causes his sun to rise on the evil and the good, and sends rain on the righteous and unrighteous." Clearly, the authorial intention of the Lord in this verse and that of the Holy Spirit and human authors in the three other passages above is not to reveal scientific facts about the sun. Instead, God accommodated to the level of ancient people and used their ancient astronomy as a vessel to declare that he was Creator and Lord.

6. Falling Stars & Rolling Up of the Heavens

Everyone has heard the expression "a falling star." But today we know that this refers to a streaking meteor plunging toward the earth, and not an actual star, because just one star hitting our planet would completely annihilate us. The size and luminosity of meteors is like that of stars. For ancient people to think that stars could fall to earth is quite reasonable since they look like tiny specks in the night sky, and the occasional sighting of a streaking meteor gives the impression a star has dislodged from the firmament and fallen to earth.

Jesus accommodated and used this ancient astronomical concept in his teaching about the end of time. In Matthew 24:29, he states that "the stars will fall from the sky, and the heavenly bodies will be shaken." This verse makes sense from an ancient phenomenological perspective. The shaking of the firmament dislodges the miniscule stars, and they then drop to earth. In another end times passage, Revelation 6:13 records that "the stars in the sky fell to earth, as figs drop from a fig tree when shaken by a strong wind." And in Revelation 12:4, the apostle John envisions a "red dragon" whose "tail swept a third of the stars out of the sky and flung them to the earth."

As we have seen earlier, Scripture compares the dome of heaven to the canopy of a tent. For example, Psalm 104:2 states, "He [God] stretches out the heavens like a tent" (also Ps. 19:4-5; Is. 40:22). It was logical for ancient people to believe that the heavens would be rolled up like a tent canopy at the end of time. The Bible also describes the dismantling of the heavens as the rolling up of a scroll. Isaiah 34:4 asserts that "the heavens rolled up like a scroll," and so too Revelation 6:14, "The heavens receded like a scroll being rolled up." If heaven refers to outer space, then how can it be rolled up? The answer is found only in the context of a 3-tier universe. The firmament of heaven is a physical structure, and therefore, it can be rolled up like a scroll or tent canopy.

These prophetic passages above picture the end of the world as the dismantling of the heavens in a 3-tier universe. With this being the case, the actual physical changes in the cosmos at the Second Coming of Jesus will not happen as described in Scripture. This is not a problem for us as Bible-believing Christians, because the Holy Spirit accommodated by allowing the inspired biblical writers to use their ancient astronomy. In other words, descriptions of God's action in ending the present world are filtered through ancient astronomical concepts in the same way that the creation of the world in Genesis 1 is accommodated through an ancient *de novo* understanding of origins.

We may not know the exact details of how God will dismantle the universe at the end of time, but the message of faith in the Word of God is exceedingly clear. The world as we know it will come to an end when Jesus returns again. We can be certain this will happen.

7. The Lower & Upper Heavens

As twenty-first century Christians, we often think of heaven as being outside the universe in a different dimension. However, in the Bible, heaven refers to a physical region in this present world. The Hebrew word for "heaven" is *shāmayim*. As you can see, it has the dual ending *-ayim* which is used to designate things in nature that occur in pairs. This term appears about 400 times in the Old Testament, and it is always in the dual and never in the singular or the plural. Therefore, the noun *shāmayim* indicates that there are two heavens in the physical world—the lower heavens and the upper heavens.

The lower heavens include the firmament. For example, on the second day of creation, the Creator "called the firmament 'heavens' " (Gen. 1:7). The lower heavens also consist of the air or atmosphere as seen with the phrases "the birds of the air [*shāmayim*]" (Gen. 2:19-20; NIV 1982), and "the clouds of heaven" (Dan. 7:13). The lower heavens also extend to the heavenly sea and the "waters in the heavens" (Jer. 10:13).

The upper heavens are where God and his angels dwell. Understood from an ancient phenomenological perspective, it is a real physical region just overhead and above the heavenly sea. For example, Exodus 24:9-10 describes Moses and the leaders of Israel going up Mount Sinai to meet God. "Moses, Aaron, Nadab and Abihu, and the seventy elders of Israel went up and saw the God of Israel. Under his feet was something like a pavement made of lapis lazuli, as bright blue as the sky" (cf. Ezek. 1:26-28; 10:1). The "pavement" refers to the base of the divine dwelling set upon the heavenly sea. For the ancient Hebrews, God is just above them.

In addition, Psalm 104:2-3 describes, "He [God] stretches out the heavens like a tent and lays the beams of his upper chambers on their waters." The "upper chambers" refers to God's heavenly dwelling. In this context, the prayer to the Lord in Deuteronomy 26:15 makes perfect sense. "Look down from heaven, your holy dwelling place, and bless your people Israel." Occasionally the Old Testament qualifies the upper heavens as *shāmayim shāmayim*, which literally means "heavens of heavens." Nehemiah 9:6 states that God "made heaven [lower heavens]" and "the heaven of the heavens [upper heavens]" (NRSV).

In the New Testament, the Greek noun *ouranos* is translated as "heaven," "sky," and "air." In the lower heavens, this term can mean the atmosphere in the phrases "the birds of the air [*ouranos*]" (Matt. 6:26) and "the clouds of heaven" (Mk. 14:62). *Ouranos* also refers to the firmament since it can be opened (Jn. 1:51), shaken (Matt. 24:29), and rolled up (Rev. 6:14). It is implied in the term "the stars of the sky [*ouranos*]" (Phil. 2:15), since stars are embedded in the firmament according to ancient astronomy. In the upper heavens, *ouranos* is where God and his angels reside (Mk. 12:25; Jn. 6:38). Mark 1:10-11 states, "Just as Jesus was coming up out of the water, he saw heaven being torn open and the Spirit descending on him like a dove. And a voice came from heaven: 'You are my Son, whom I love; with you I am well pleased.'"

Figure 15-1 (p. 109) identifies the lower heavens and the upper heavens in the 3-tier universe. Clearly, the authorial intention of the Holy Spirit in Scripture was not to reveal scientific facts about the structure and operation of the astronomical world. Instead, God accommodated and allowed an ancient astronomy to reveal that he is the Creator and Lord of the heavens. In particular, the Bible affirms the reality of a region where God and angels dwell. Like the underworld, the upper heaven represents a real place, even though the Word of God does not reveal the actual location of these two regions. My *speculation* is that the realm of dead human souls (spirits) and the residence of the Lord are outside the present physical creation in another realm or dimension.

Conclusion
The Tent Model/Metaphor of the Universe

The use of familiar objects and processes as models and metaphors to describe and to explain the structure and operation of the world is a common practice in science today. For example, in a recent scientific paper on dentition development, I refer to an "odontogenic field" in which teeth initiate in a jaw like plants sown within the boundaries of a farmer's field (Greek *odonto-*: tooth; *-genic*: origin).[3] Ancient people also employed models and metaphors to understand nature, such as the perceived similarity between agriculture and reproductive biology (e.g., seeds and barrenness).

In Hermeneutical Principles 15 and 16, we noted that biblical authors compared the structure of the universe to a tent. As a conclusion to these two interpretive principles, let's revisit these tent passages and read them at the same time.

> He [God] sits enthroned above the circle of the earth, and its people are like grasshoppers. He stretches out the heavens like a canopy, and spreads them out like a tent to live in.
>
> Isaiah 40:22
>
> He [God] stretches out the heavens like a tent and lays the beams of his upper chambers on their waters.
>
> Psalm 104:2-3
>
> In the heavens God has pitched a tent for the sun. It is like a bridegroom coming out of his chamber, like a champion rejoicing to run his course. It rises at one end of the heavens and makes its circuit to the other.
>
> Psalm 19:4-6

Taken together, these passages in Scripture paint a *coherent* picture of the physical world as depicted in Figure 15-1 (p. 109). Viewed from an ancient phenomenological perspective, the domed "canopy" of the firmament encloses the flat "circle of the earth," which implies the earth is bordered by a circumferential sea. The firmament supports the waters in heaven, and God's "upper chambers" with his throne is set on this body of water. And comparing the sun to the joyful stride of a "bridegroom" or "champion" refers to the literal movement of the sun across the firmament in a "circuit" from one "end" of the firmament at the horizon to the other. The use of a tent as a model and metaphor in these biblical verses clearly indicates that the Holy Spirit accommodated and allowed the inspired authors to include their ancient geography and ancient astronomy in the Word of God.

The similarities are very clear between the tent model/metaphor of the universe in the Bible and the understanding of the structure and operation of the world in ancient Egypt and ancient Mesopotamia (Fig. 15-1, p. 109; Figs. 14-3 to 14-6, pp. 102-105). However, there are radical theological differences. The Egyptian and Mesopotamian depictions of nature contain multiple gods and a pagan spirituality. In sharp contrast, Scripture frees the universe of these gods. The Word of God strips the earth and heaven of their so-called "divinity" and turns them into "very good" creations (Gen. 1:31). As a result, the biblical understanding of the universe places the God of Scripture as the only Creator and Lord over the entire world.

HERMENEUTICAL PRINCIPLE 17

The Accommodation of God's Creative Action in Origins

As I mentioned earlier, I want you to be fully aware that I had an agenda in Hermeneutical Principles 15 and 16 by presenting the 3-tier universe in Scripture. My goal was to make you feel comfortable with the idea that the Bible features an ancient geography and ancient astronomy. Because, if the Word of God features a world with three tiers, then the implications for the origins debate are quite significant. More specifically, acknowledging this ancient science will impact our understanding of God's creative action in the biblical accounts of creation.

For many students in my college course on science and religion, this hermeneutical principle is the most challenging of all. Let me introduce it by asking three questions.

First, does the Bible reveal the actual structure and operation of the heavens and the earth? The answer is "no." Scripture has a 3-tier universe as shown in Figure 15-1 (p. 109). According to this ancient science, the heavens are made up of a solid domed firmament that upholds a heavenly sea. God's dwelling rests upon this body of water in heaven. The sun, moon, and stars are positioned in the firmament. The earth is flat and surrounded by a flat circumferential sea. The underworld is below the surface of the earth. And the sun moves across the dome of heaven every day, and the earth is stationary and never moves. But as everyone knows, this is not the structure of the world or how it operates.

Second, does the Bible reveal how God actually created the universe? Again, the answer is "no." Let's focus on the creation of the heavens. Genesis 1:6-8 states that God created the firmament on the

THE BIBLE & ANCIENT SCIENCE

second creation day and used it to separate the heavenly "waters above" from the earthly "waters below." Genesis 1:14-19 then asserts that on the fourth day the Creator made the sun, moon, and stars and that he placed these heavenly bodies in the firmament. However, we all know that there is no solid dome overhead. There is no heavenly sea of water above the earth. And there are no astronomical bodies set in a solid firmament. Therefore, Genesis 1:6-8 and 14-19 are not an account of actual events in how the heavens were created.

To be more precise, the origin of the heavens on creation days two and four begin with the clause, "And God said, 'Let there be . . .'" (Gen. 1:6, 14). In these verses, the Bible presents the commands of the Creator in his making of the heavens. And to intensify this situation, these are *God's very words in the Word of God*. However, it is quite obvious that the Lord never actually said these commands because he never created the heavens of a 3-tier universe. It never really happened.

Third, and there is nothing wrong with us asking the question, did God lie in the Bible regarding the structure, operation, and origin of the heavens and the earth? My answer is an emphatic "NO!!!" But do you feel the tension? You should if you are a Bible-believing Christian. Rarely is this ancient science in Scripture presented in our churches or Sunday schools. So how do we explain and justify statements in Genesis 1 about God creating a 3-tier universe?

The answer is quite simple, and you already know it—God accommodated. During the process of biblical revelation, the Holy Spirit came down to the level of the ancient human author of Genesis 1 and allowed him to use his ancient understanding of the structure, operation, and origin of the world. According to ancient people, the Creator made a 3-tier universe and he created it *de novo* (quick and complete). They had no idea that the cosmos began with the Big Bang 14 billion years ago, and that it evolved into the incalculable number of galaxies, stars, planets, and moons existing in the heavens today. Not having modern telescopes, ancient individuals had a very narrow scope of cognitive competence. And had we lived in their day, we would have come to the very same conclusion—God created a world with three tiers quickly and completely.

We can now define Hermeneutical Principle 17, the accommodation of God's creative action in origins. The acts of the Creator in the biblical creation accounts are filtered through the ancient understanding of *de novo* creation. In other words, the Holy Spirit accommodated and permitted the inspired writer of Scripture to employ the origins science-of-the-day as a vessel to deliver the central inerrant truth that the God of the Bible is the Creator of the entire world.

The accommodation of divine action in origins has two important implications for the modern origins debate. First, the Bible does not reveal how God actually created the world. Consequently, we need to explore the natural world and do science in order to discover his creative method. Second, since Scripture features an ancient understanding of the origin of the heavens and the earth, it is only reasonable to suggest that the Word of God also has an ancient view of the creation of living organisms . . . including men and women.

Historical Criticism & the Biblical Creation Accounts

As we noted in Hermeneutical Principles 14-16, an examination of the geography and astronomy held by ancient nations surrounding the biblical authors offers valuable insights into understanding the 3-tier universe in the Bible. Let's now look at two accounts of origins from ancient Mesopotamia. Notably, divine creative action in these stories is similar to the biblical creation accounts in that the god/s employ *de novo* creation—the inanimate universe and living organisms are created rapidly and fully formed.

Bilingual Creation of the World by Marduk

This creation story features the main Mesopotamian creator god Marduk. It dates to about the sixth century BC, but originates much earlier, somewhere between 3000-2000 BC.[1] The opening scene in this account depicts a pre-creative state dominated by water.

> All the lands were sea . . .
> Marduk constructed a raft on the waters;
> He created dirt and piled it on the raft.
> In order to settle the gods in the dwelling pleasing to them,

He created humankind.
Aruru [goddess of child birth] created the seed of human-
   kind with him.
He [Marduk] created the wild animals and all the animals
   of the steppe.
He created the Tigris and the Euphrates and set them in
   place,
Giving them a favorable name.
He created the grass, the rush of the marsh, the reed, and
   the woods;
He created the green herb of the field,
The lands, marshes, and canebrakes,
The cow and her young, the calf, the ewe and her lamb,
   the sheep of the fold;
The orchards and forests,
The wild sheep, the ibex (words missing) to them.
He made an embankment along the sea.
(words missing) dried up (?) the swamp.
He cause to appear (words missing)
He creat[ed the reed], he created the tree;
[Bricks he laid, the br]ick mold he built;
[Houses he built,] cities he built;
[Cities he made,] living creatures he placed [therein];
[Nippur he built], Ekur he built;
[Uruk he built, Eann]a he built.[2]

A number of striking similarities appear between Genesis 1 and *Bilingual Creation of the World*. Both accounts begin with a pre-creative state in which there was no dry land and water completely dominated the world. The Mesopotamian account states, "All the lands were sea;" and Genesis 1:2 asserts, "Now the earth was formless and empty, darkness was over the surface of the deep, and the Spirit of God was hovering over the waters." The notion of a watery state prior to the creation of the world is an ancient scientific concept that is commonly found in ancient Near Eastern accounts of origins.

Divine action in the first chapter of Scripture and this Mesopotamian story of origins is through *de novo* creation. Inanimate objects and living organisms are created quickly and completely formed. God in Genesis 1:9 commands water to be gathered in one place and dry land to appear; Marduk makes dry ground by creating dirt and placing it on a raft that floats on the pre-creative waters. This Mesopotamian god also forms the Tigris and Euphrates Rivers, and these rivers are mentioned in Genesis 2:14.

In these two creation accounts, God and Marduk make plants and animals. In Genesis 1:11, the Creator creates vegetation, including seed-bearing plants and fruit-bearing trees. Marduk also forms plant life typically found in Mesopotamia—grass, reeds, rushes, herbs, canebrakes, orchards, and trees. Genesis 1:24 states that God makes livestock, wild animals, and creatures that move along the ground. In the Mesopotamian account, Marduk creates domesticated animals like cows, ewes, and sheep as well as wild animals such as wild sheep and ibexes (wild goats).

Notably in the *Bilingual Creation of the World*, the god Marduk creates humans with the aid of the mother goddess Aruru. More specifically, she makes the "seed of humankind," reflecting the ancient reproductive notion of preformatism (1-seed model). In many Mesopotamian creation accounts, the reason people are created is to be slaves that relieve the gods of work. This is subtly implied in the statement, "In order to settle the gods in the dwelling pleasing to them, he [Marduk] created humankind."

However, in sharp contrast, Genesis 1:27 states, "So God created mankind in his own Image, in the Image of God he created him; male and female he created them." We are not slaves of the gods, but instead we are special and unique in that we are the only creatures who bear God's Image. Moreover, the Creator has made us to be rulers over his creation. In Genesis 1:28, God commands humans, "Rule over the fish in the sea and the birds in the sky and over every living creature that moves on the ground."

In fact, studies in historical criticism reveal that the term "Image of God" was used in the ancient Near East to designate human kings as representatives of the gods on earth.³ But in a radical shift away from its common usage in the ANE, the biblical author under the inspiration of the Holy Spirit recasts this category and democratizes it. Genesis 1 declares that the Lord has made all of us to be kings and queens to rule over his entire creation. Indeed, that's quite an honor!

Enuma Elish

The *Enuma Elish* is another ancient Mesopotamian creation story featuring the creator god Marduk. Written around the fourteenth to twelfth centuries BC, the opening scene has only pre-creative waters in existence.[4]

> When on high the heaven had not been mentioned,
> Firm ground below had not been called by name,
> Nothing but primeval Apsu [fresh water], their begetter,
> And Mummu-Tiamat [sea water], she bore them all,
> Their waters commingling as a single body [pre-creative waters];
> No reed hut had been matted, no marsh land had appeared,
> When no gods whatever had been brought into being,
> Uncalled by name, their destinies undetermined,
> Then it was that the gods were formed within them [pre-creative waters].[5]

According to Mesopotamian mythology, the male god Apsu and female goddess Tiamat were the parents of the first generation of gods and goddesses. In this account, these two divine beings are personified as "fresh water" and "sea water," respectively.

The opening scene of the *Enuma Elish* shares both similarities and differences with Genesis 1:2. Each features a pre-creative watery state with no mention of when it was created, if indeed it was created.[6] As we noted above, this notion was a common idea in the ancient Near East. However, Genesis 1:2 is radically different from this Mesopotamian creation account. There is no hint that these pre-creative waters are divine, and they are merely named "the deep" and "the waters." Most importantly, the God of the Bible does not arise out of these waters. And there is no mention of other gods. Genesis 1:2 places the Spirit of God in complete and total control over the pre-creative watery state. The God of Genesis 1 is the only God and Creator of the world.

The *Enuma Elish* is also marked by violent encounters between the gods. Marduk kills Tiamat and rises to become the chief god, and "lord of all the gods of heaven and the underworld."[7] He then takes her body to create the heaven and the earth.

> He [Marduk] split her [Tiamat] in half like a dried fish.
> Then he set half of her up and made the Heavens as a roof.
> He stretched out a skin and assigned a guard.
> He ordered them not to let her waters [in heaven] escape. . .[8]
> He created the spaces and fashioned firm ground . . .[9]
> He set up her head, heaped up dirt.
> Then he opened up the spring, it became saturated with water.
> The he opened the Euphrates and Tigris in her eyes.
> He plugged her nostrils, left . . . [?] . . . behind.
> He heaped up the "distant" mountains on her breast.[10]

The ancient scientific notion of a 3-tier universe appears in the *Enuma Elish* with reference to the (1) "heaven/s," (2) "firm ground," and (3) "underworld." Marduk uses half of Tiamat's body to make the "roof" of heaven. Similarly, Genesis 1:6-8 describes the creation of the "firmament." The *Enuma Elish* also refers to "waters" in heaven being held back by "a skin" and "a guard." This ancient idea appears in Genesis 1:7 with the firmament supporting the heavenly "waters above." Marduk then makes the earth with the other half of Tiamat's body. He uses her eyes as sources of the Euphrates and Tigris rivers. But Genesis 2:14 only refers to these rivers and there is no mention they derive from divine body parts. More importantly, the God of Genesis 1 creates the world with utter ease through spoken commands. There is no hint of any other gods or any opposition to the biblical Creator.

After the creation of the heavens and the earth in *Enuma Elish*, Marduk "made up his mind to perform miracles."[11] He fills the world through *de novo* creative acts and also establishes natural processes to provide nourishment for living organisms. In particular, this account focusses on the creation of agriculture. Marduk is the "bestower of cultivation" and "creator of grain and herbs who causes vegetation to sprout."[12] He also "establishes seed-rows," "forms fine plow land in the steppe," "furnishes millet," "causes barley to appear," "furnishes the seed of the land," and "causes rich rains over the wide earth, provides vegetation."[13] Though not explicitly stated, Marduk also created animals since he is the "creator of all."[14] He "provides grazing and drinking places" and "enriches their stalls."[15]

In a similar way, after God forms dry land in Genesis 1, he creates vegetation *de novo* using natural processes on the third day of creation. Verse 11 states, "Then God said, 'Let the land produce vegetation: seed-bearing plants and trees on the land that bear fruit with seed in it, according to their various kinds.'" And on the sixth day of creation, the Creator offers plant life as sustenance to humans and animals. Verses 29 and 30 record, "Then God said, 'I give you [humans] every seed-bearing plant on the face of the whole earth and every tree that has fruit with seed in it. They will be yours for food. And to all the beasts of the earth and all the birds of the sky and all the creatures that move along the ground—everything that has the breath of life in it—I give every green plant for food.'"

The origin of humans in the *Enuma Elish* is through *de novo* creation, but once again violence between the gods characterizes this Mesopotamian story. After killing Tiamat and using her body to make the heaven and the earth, Marduk decides, "Let me create primeval man. The work of the gods shall be imposed (on him), and so they shall have leisure."[16] Tiamat's battle commander Qingu is captured and executed, and Marduk "created mankind from his blood, [and then] imposed the toil of the gods (on mankind) and released the gods from it."[17]

The creation of men and women in Genesis 1 and 2 is also *de novo*. Humankind is made quickly on the sixth day of creation in Genesis 1. The Lord in Genesis 2 fashions Adam and Eve completely formed.

However, the reason for creating humans in the biblical creation accounts is drastically different than the *Enuma Elish*. Instead of making them slaves of the gods, God in Genesis 1 honors men and women with his Image and places them as his representatives and rulers over the world. And Genesis 2 makes it clear that the Creator is not the tyrannical taskmaster of the Mesopotamians. Instead, the Lord creates a garden paradise for Adam and Eve and furnishes all their needs. They even enjoy a personal relationship with him.

Historical criticism opens our understanding to the scientific concepts and mindset of ancient Near Eastern people, including the inspired writers of the Bible. The *Bilingual Creation of the World by Marduk* and the *Enuma Elish* both demonstrate that *de novo* creation was the origins science-of-the-day. The inanimate world and living organisms were created rapidly and fully developed. This type of divine creative action also appears in Genesis 1 and 2. However, the ancient view of origins in these opening chapters in Scripture is incidental to the timeless messages of faith. Most importantly, the biblical accounts of origins reveal the inerrant spiritual truths that we are God's special creation and that he very much cares for us.

HERMENEUTICAL PRINCIPLE 18

De Novo Creation of Living Organisms: Ancient Biology

In previous hermeneutical principles, we discovered that ancient biology appears in Scripture. Ancient taxonomy considers bats to be birds (Lev. 11:13-19) and rabbits as ruminants that chew the cud (Lev. 11:6). Ancient botany asserts that the mustard seed was the smallest seed on earth (Mk. 4:26-29) and that seeds die before they germinate (Jn. 12:23-24). Ancient reproductive biology (preformatism or 1-seed model) assumed that only men had reproductive seed (Lev. 15:32) and women were like fields that nurtured male seed (Num. 5:28). In addition, only women were believed to be barren and the cause of infertility (Gen. 11:30).

By viewing living organisms through an ancient phenomenological perspective, the biblical writers and their readers also believed plants and animals were immutable and unchanging. They never observed one kind of living organism transform into another kind of living organism. For example, ancient people would only have seen that a goat gives birth to a goat, which gives birth to goat, which gives birth to a goat, etc. So it is understandable why the Genesis 1 creation account states that God created plants and animals "according to their/its kinds" ten times (v. 11, 12 twice, 21 twice, 24 twice, 25 thrice). Biological immutability was the science-of-the-day.

In order to understand the creation of each kind of living organism, ancient individuals would have simply reversed their observation that one kind of creature arises from the same kind of creature backward in time. To use the example of goats, they would have reasoned that a goat today was birthed from an earlier goat in the past, which was birthed from an even earlier goat in the more distant past, etc. And since ancient people believed that goats were immutable, they logically concluded that at the beginning of the world there must have been some original goats that God had created *de novo*—quickly and completely mature.

The intellectual process of reversing a series of events back in time, like goats birthing goats, is called "retrojection." The Latin adverb *retro* means "backwards" and *jactare* is the verb "to cast" or "throw." This type of thinking is used today in crime scene investigations (CSI). Police collect evidence present at the scene and then cast it back in time to recreate the criminal acts. The biblical authors were no different than us in trying to understand the past. In their re-creation of the events in the origin of living organisms, they retrojected their observation that plants and animals only give rise to their own kinds, and came to the reasonable conclusion that each kind of creature was originally made rapidly and fully-formed.

This ancient conceptualization of the origin of living creatures is what we find in Genesis 1. God creates plants *de novo* on creation day three, sea creatures and flying creatures *de novo* on day five, and land animals and humans *de novo* on day six. In biblical times, the fossil record had not yet been discovered, and there was no reason for ancient people like the inspired writers of Scripture and their readers to believe that life evolved over billions of years. *De novo* creation was the best origins science-of-the-day in the ancient world.

Now here is where it gets quite challenging. Each day of creation in Genesis 1 that deals with the origin of living organisms begins with a divine command. The third creation day opens, "*God said*, 'Let the land produce vegetation'" (v. 11). On the fifth day, "*God said*, 'Let the water teem with living creatures, and let birds fly above the earth'" (v. 20). The sixth creation day begins, "*God said*, 'Let the land produce living creatures . . . the livestock, the creatures that move along the ground, and wild animals'" (v. 24). Later on the sixth day, "*God said*, 'Let us make mankind'" (v. 26; all my italics).

And this is the incisive question we need to ask: Is the accommodation of God's creative action in the origin of living organisms in Genesis

1 similar to his *de novo* acts in creating the heavens and the earth in this biblical creation account? My answer is a definite "yes."

In the same way that the Holy Spirit accommodated God's creative action in the *de novo* creation of the firmament, waters above, sun, moon, stars, and dry land, he also allowed the human biblical writer to use the ancient science of biological origins—the *de novo* creation of living organisms. Once again, we have divine commands in Genesis 1 with *God's very words in the Word of God* that never actually happened.

To be sure, this is quite a shocking statement to most Bible-believing Christians. However, this situation is easily resolved. The *de novo* creation of plants, animals, and humans is an incidental ancient vessel that transports the inerrant spiritual truth that the God of the Bible is the Creator of all living creatures, including men and women. Therefore, if we want to discover how the Lord made these creatures and us, we need to practice science and explore his creation.

Historical Criticism & the Biblical Creation Accounts

As we have noted several times, historical criticism helps us to better understand the Bible by outlining the setting in which it was written and read. In order to shed light on the creation of living organisms in Scripture, let's consider accounts of human origins from ancient Near Eastern nations surrounding the biblical authors. These ancient stories reveal that there were two basic methods of creation. The first is a plant-like mechanism in which people sprout out of the earth (yes, you are reading that correctly!). The second is a craftsman-like technique where humans are formed from earth or clay.

To comprehend the first creative method, we need to remember that ancient people accepted the ancient biological concept of preformatism (1-seed model). They believed that a miniscule person was inside the reproductive seed of men. With this being the science-of-the-day, the *Assur Bilingual Creation Story* records that the gods plant the seeds of humans into the earth and people later "sprout from the ground like barley."[1] Similarly, the *Hymn to E'engura* asserts "humans broke through the earth's surface like plants."[2] And in the *Song of the Pickaxe*, a god uses a hoe-like axe to till the ground "so that the seed from which people grew could sprout from the field."[3]

This plant-like mechanism appears to be the creative method used in Genesis 1:24 when God commands, "Let the land produce living creatures according to their kinds: the livestock, the creatures that move along the ground, and the wild animals, each according to its kind." The Hebrew verb translated as "produce" in this verse is *yāṣā'* and it is the same verb found in Genesis 1:12 in which "the land produced vegetation." This ancient creative mechanism for the origin of land creatures certainly strikes us as hard to believe. However, Genesis 1 clearly implies that land animals as well as land plants were created from seeds that sprouted out of the earth.

The craftsman-like method for fashioning humans also appears in various ancient Near Eastern creation accounts. In the *Epic of Atrahasis*, a goddess mixes clay with the blood of a slaughtered god to form seven men and seven women.[4] The story of *Enki and Ninmah* recounts that an intoxicated female divine being uses earth to make seven defective human beings.[5] In the *Epic of Gilgamesh*, a goddess creates a man from a pinch of clay.[6] And according to the Egyptians, the god Khnum creates people from clay by fashioning each on a potter's wheel, and then the goddess Isis gives life to them.[7] Figure 18-1 shows this ancient understanding of the creation of humans.[8]

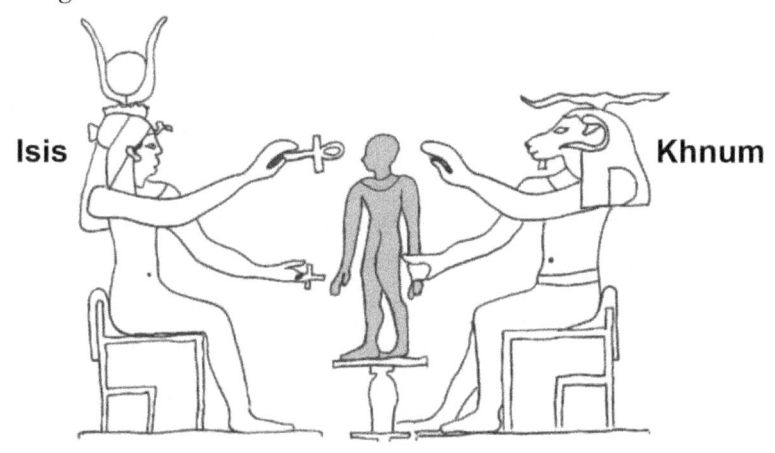

Figure 18-1. Ancient Egyptian Creation of Humans

Clearly, these ancient Near Eastern examples of the *de novo* creation of humans are similar to Genesis 2:7, in which the Lord God acts like a craftsman and "formed a man from the dust of the ground." Notably, the word translated as "formed" in this verse is the Hebrew verb *yāṣar* and it is the root of the word "potter." For example, this noun appears in Isaiah 29:16 and 64:8 where God is depicted as a divine potter who forms man in his hands from clay. In addition, Genesis 2:7 states that the Lord God "breathed into his [the man's] nostrils the breath of life, and the man became a living being." This is like the Egyptian goddess Isis who gives life to the clay humans made on the potter's wheel by the god Khnum.

Of course, it is certainly challenging for most Christians to grasp that the Bible includes an ancient biological understanding of the origin of living organisms, including humans. The ancient notion of *de novo* creation has significant implications for the origins debate. According to young earth creation and progressive creation, God created plants and animals through dramatic miraculous acts, as recorded in Genesis 1 and 2. However, these anti-evolutionists fail to recognize and respect that God's creative action is filtered and accommodated through the ancient scientific notion of *de novo* creation. And as we have noted above, these events never happened. Young earth creationists and progressive creationists do not realize that their view of origins is ultimately based on an ancient biology.

The Message-Incident Principle helps us understand that the primary purpose of the Bible is to reveal inerrant spiritual truths. The main message in Genesis 1 and 2 is not *how* God actually created living organisms. Instead, Scripture declares *who* created plants, animals, and humans—the God of the Bible. And in particular, of all the living creatures made by the Creator, Genesis 1:26-27 reveals that men and women are unique and special because we alone have been created in the Image of God.

To conclude our examination of the ancient geography, ancient astronomy, and ancient biology in Scripture that was presented in Hermeneutical Principles 15-18, we can now answer questions that were asked about historical criticism in Hermeneutical Principle 14. To use George Eldon Ladd's aphorism, are the accounts of origins in Genesis 1 and 2 "the Word of God given in the words of men in history?"[9] Or employing John Walton's insight, were the biblical creation accounts "written *for* us," but "not written *to* us?"[10] And to be more precise, "Are these opening chapters in Genesis historically conditioned?" My answer to these three questions is a firm "yes." Historical criticism and the study of the literature and images from ancient Near Eastern nations surrounding the inspired biblical writers assist us to identify the incidental ancient view of origins that transport the inerrant spiritual truths in Genesis 1 and 2.

Excursus
The God-of-the-Gaps

Now that you are familiar with the accommodation of God's creative action in origins, I can introduce a term that often appears in discussions on origins—the God-of-the-gaps. According to this notion, there are "gaps" in the continuum of natural processes, and these "breaks" in nature indicate places where God has miraculously intervened in the world. This view of divine action usually carries a negative nuance and pictures God as a meddler who tinkers about in the physical world sporadically during the origin and operation of the universe and life. However, it must be emphasized that the Creator can act in any way he wants and at any time he wants, including through dramatic interventions in nature.

It is necessary to qualify that the God-of-the-gaps is limited to divine action in the origin and operation of world. This notion does not extend to the Lord's miraculous acts in the personal lives of men and women. Therefore, young earth creation and progressive creation embrace a God-of-the-gaps view for the origin of living organisms. In contrast, evolutionary creation rejects this approach. But these three Christian views of origins fully believe that the Lord acts miraculously with people (Appendix 1).

THE BIBLE & ANCIENT SCIENCE

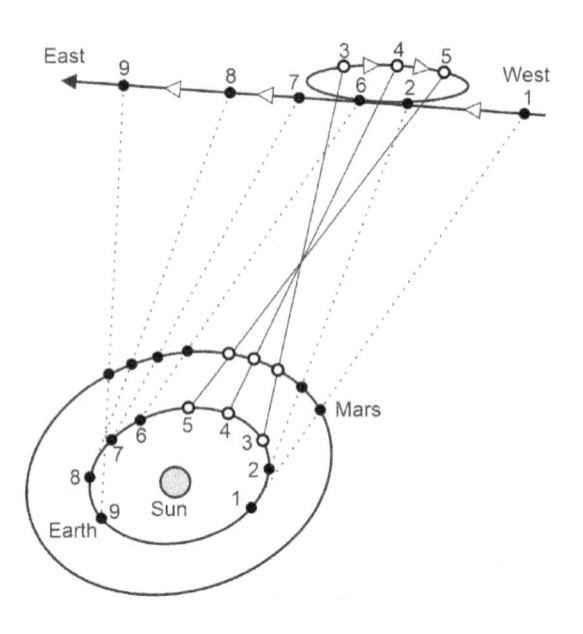

Figure 18-2. Retrograde Planetary Motion

One of the best ways to explain the God-of-the-gaps is to use retrograde planetary motion. From our perspective on earth, planets like Mars travel mostly from west-to-east in the sky. But they also make a brief east-to-west loop, which is termed "retrograde motion." As we noted above, the Latin *retrō* is the adverb "backwards." The verb *gradior* means "to go" and "step." Today, we know that this "movement" is only a visual effect caused by the earth traveling past a planet as shown in Figure 18-2.[11]

In the past, astronomers believed that the earth was at the middle of the universe (geocentricism) and that retrograde planetary motion was real, and not just an appearance of motion. In fact, some religious individuals believed that God intervened and actually caused planets to make these brief backward loops. For example, theologian Martin Luther claimed, "The retrograde motion of the planets also is a work of God, created through His Word. This work belongs to God Himself and is too great to be assigned to the angels."[12] But as we have noted in Hermeneutical Principle 4, Luther was a geocentrist (cover of the book), and as a result, he did not understand planetary motion as we do today.

Martin Luther demonstrates the problem with the God-of-the-gaps. It is not that there is a *gap in nature* where God actually enters the world and causes a planet to make a backward loop. Rather, Luther had a *gap in knowledge*. Therefore, he appealed to a dramatic miraculous intervention by God in order to explain the retrograde planetary motion he saw in the heavens.

Once again, I must underline that the God-of-the-gaps understanding of divine action is reasonable. If there are gaps in the continuum of natural processes, then science will identify them, and over time they will "widen" with further research. That is, as scientists explore a true gap in nature, physical evidence will increase and point away from any natural processes accounting for the origin or operation of a feature in the world. Should this ever be the case, it would be very reasonable to conclude that a divine intervention had occurred in the past.

However, there is a clear trend in the history of science. The God-of-the-gaps view of divine action has repeatedly and consistently failed. Instead of the gaps in nature getting "wider" with the advance of science, they have always been "closed" or filled by the growing body of scientific knowledge. History reveals that these gaps have always been gaps of information, and not actual gaps in the natural world indicative of the intervening hand of God.

We can now fully appreciate the origin of living organisms promoted by young earth creation and progressive creation. First and foremost, these anti-evolutionary positions are entrenched in scientific concordism. They assume that the *de novo* creative acts in Genesis 1 and 2 are a record of actual events in the origin of plants, animals, and humans. But as we have seen, God's creative action in these biblical chapters is filtered and accommodated through an ancient biology. Therefore, these divine acts of creation in Scripture dealing with the origin of living creatures never actually happened.

Second, the concordist hermeneutic of young earth creationists and progressive creationists forces them to accept a God-of-the-gaps understanding of the origin of living creatures. As a consequence, they completely disrespect and disregard the overwhelming evidence for biological evolution and the fact that 98% of scientists today accept that "humans and other living things have evolved over time."[13] In this way, these anti-evolutionists have a gap in their knowledge regarding biological origins and they assume there are actual gaps in nature requiring dramatic *de novo* action by the Creator to make plants, animals, and us. But like the ancient geocentric understanding of retrograde planetary motion, young earth creationists and progressive creationists attribute to God miraculous interventions in biological origins that never really happened.

HERMENEUTICAL PRINCIPLE 19

Does Conservative Christianity Require Scientific Concordism?

We have now come to a point in this book where we can ask some critical questions. What distinguishes traditional conservative Christianity from modern liberal Christianity? Will our interpretation of Scripture regarding the structure, operation, and especially the origin of the world, be a determining factor? For example, are scientific concordists like young earth creationists and progressive creationists the only conservative Christians? And are non-concordists, such as evolutionary creationists, to be considered as liberal Christians? Or to be more precise, do we need to accept scientific concordism in order to be a conservative Christian? To answer these questions, let's turn again to astronomy and examine the views of three of the most important theologians in church history—St. Augustine, Martin Luther, and John Calvin.

In his AD 415 book *The Literal Meaning of Genesis*, St. Augustine reveals that in his day there was an intense debate within academic circles regarding the relationship between the Bible and astronomy. He writes, "It is also frequently asked what our belief must be about the form and shape of heaven *according to Sacred Scripture*. Many scholars engage in lengthy discussions on these matters" (my italics).[1] Clearly, scientific concordism played a significant role in hermeneutics during that time. Augustine observed that there were three different ways to align the Bible with science.

Some academics accepted a form of geocentricism in which "heaven is like a sphere and the earth is enclosed by it and suspended in the middle of the universe" (Fig. 9-1, p. 66).[2] Others argued that if heaven "is not a sphere, it is a vault on that side on which it covers the earth."[3] This describes the 3-tier universe with a domed heaven and a flat earth (Fig. 15-1, p. 109). And there were scholars who believed that "heaven [is] like a disk above the earth [and] covers over it on one side."[4] This view contended that the sky was flat and circular. Augustine understood these astronomical theories as credible possibilities and did not seem to have embraced one in particular.[5]

St. Augustine also recognized "a great deal of subtle and learned enquiry" concerning whether the firmament is "in reality stationary or moving."[6] This naturally led to the question, "If it is stationary, how do the heavenly bodies that are thought to be fixed in it [e.g., stars attached to the firmament] travel from east to west?"[7] For many scholars the well-known observation that all the stars moved together in unison was explained by "the picture of heaven turning either like a sphere . . . or like a disk."[8] But Augustine acknowledged that "if the [scientific] evidence shows that the heavens actually are immovable, the motion of the stars will not be a hindrance to our acceptance of this fact."[9]

However, Augustine believed the Bible did offer some undeniable scientific facts regarding the physical world. In appealing to the second day of creation in Genesis 1:6-8, he argues, "Bear in mind that the term 'firmament' does not compel us to imagine a stationary heaven: we may understand this name as given to indicate not that it is motionless but that it is solid and that it constitutes an impassable boundary between the waters above and the waters below."[10] It is obvious that St. Augustine was a scientific concordist. Despite the scholarly debate over the various views about the shape of the firmament, he certainly accepted that the firmament was a real solid structure in heaven, with stars embedded in it, whether they were fixed or in motion. By referring to the "waters above" mentioned in Genesis 1:7, Augustine demonstrates that he believed in the existence of the heavenly sea above the firmament.

As we have seen previously, protestant reformer Martin Luther accepted geocentrism as depicted in the diagram of the universe that appears across from Genesis 1 in his 1534 German translation of the Bible (cover of this book). Geocentricism was the science-of-the day in the sixteenth century and Luther in his 1536 *Lectures on Genesis* affirmed that "the earth is the center of the *entire* creation" (my italics).[11] Scientific concordism shaped his understanding of the structure of the world. In commenting on Genesis 1:6-8, Luther argues that God made the firmament so that "it should extend itself outward in the manner of a sphere."[12] And Luther believed that "the *primum mobile* [i.e., the firmament] revolves from east to west" and this daily rotation produces a 24-hour day on earth.[13]

Lectures on Genesis offers further evidence of Luther's concordist hermeneutics. He writes, "Scripture . . . simply says that the moon, the sun, and the stars were placed, not in individual spheres but in the firmament of the heaven (below and above which heaven are the waters) . . . The bodies of the stars, like that of the sun, are round, and they are fastened to the firmament like globes of fire."[14] As we will see below with John Calvin, some astronomers believed that there were a series of heavenly spheres with each having its own planet. But Luther rejects this view, because the Bible simply states that on the fourth day of creation God placed all the heavenly bodies "in the firmament of the heaven."

In dealing with Scripture and science, Luther delivers a blunt warning to the church. "We Christians must, therefore, be different from the philosophers [i.e., natural philosophers or scientists] in the way we think about the causes of things. And if some are beyond our comprehension (like those before us concerning the waters above the heavens), we must believe them and admit our lack of knowledge rather than wickedly deny them or presumptuously interpret them in conformity with our understanding. We must pay attention to the expression of Holy Scripture and abide in the words of the Holy Spirit."[15] Most Christians today do not believe that there is a heavenly sea just above our heads. Yet I doubt any of us think that we are acting "wickedly" or "presumptuously" for not accepting the "waters above" referred to in Genesis 1:7.

Finally, protestant reformer John Calvin also assumed that the earth was at the center of the universe. But this type of geocentricism featured a series of concentric spheres enveloping the earth. Each sphere had a planet and the innermost sphere held the moon (Fig. 9-2, p. 67). In his 1545 *Commentary on Genesis*, he confidently asserts, "We indeed are not ignorant, that the circuit of the heavens is finite, and that the earth, like a little globe, is placed in the *centre*" (my italics).[16] Calvin also believed that the firmament rotated, causing the movement of the spheres and their respective planet as well as the stars attached to the firmament. He writes, "The *primum mobile* [i.e., the firmament] rolls all the celestial spheres along with it . . . the firmament, by its own revolution, draws with it all the fixed stars."[17]

Calvin was also a scientific concordist. In his interpretation of Genesis 1:6-8, he states, "The work of the second day [of creation] is to provide an empty space around the circumference of the earth."[18] This again is indicative of a spherical earth. Notably, Calvin deals directly with the meaning of the "waters above" in this verse. He comments that the author of Genesis 1 "describes the special use of this expanse, to divide the waters from the waters from which word arises a great difficulty. For it appears opposed to common sense, and quite incredible, that there should be waters above the heaven."[19] Calvin's solution to this problem was to suggest that these heavenly waters are clouds. As he explains, "God has created the clouds, and assigned them a region above us."[20]

In a tone reminiscent of Martin Luther's warning to the church, John Calvin had harsh comments for those who accepted heliocentrism. He preached in a 1556 sermon, "We will see some who are so deranged, not only in religion but who in all things reveal their monstrous nature, that they will say that the *sun does not move*, and that *it is the earth which shifts and turns*. When we see such minds, we must indeed confess that the Devil possesses them."[21] Of course, Christians today do not believe they are mentally damaged or controlled by Satan for acknowledging the well-established scientific facts that the earth revolves around the sun and rotates on its axis. Sadly, I once believed that Christian evolutionists were scientifically incompetent and deceived by the Devil. But now I realized that comments like this are simply wrong and misguided. They do not encourage Christian unity or advance our understanding of the relationship between science and faith.

To summarize, the history of biblical interpretation reveals that three of the most important theologians in church history were scientific concordists. St. Augustine, Martin Luther, and John Calvin believed in the existence of a solid firmament above the earth. Or stated another way, the traditional and "conservative" interpretation of the structure of the heavens in Scripture for about 1500 years of church history accepted an astronomical structure that in fact *has never existed*. Today I have yet to meet a Christian who believes in the firmament. Does this mean that we are liberal Christians because we have broken away from church tradition?

The main problem with trying to align the Bible and the science-of-the-day is obvious from these passages on astronomy by these three great Christian scholars. Science has advanced and their astronomical theories have been shown to be incorrect and replaced with a more accurate understanding of nature. No one today believes that we live in either a 3-tier universe or geocentric world with a firmament overhead. Scientific progress has completely overturned the concordist interpretations of these three leading theologians, and this is more evidence for why scientific concordism fails.

The valuable lesson to be learned from history and the biblical interpretations of Augustine, Luther, and Calvin is that the distinction between conservative Christianity and liberal Christianity is not found in our understanding of the science in Scripture. Instead, it is primarily determined by our beliefs about Jesus Christ. Did God become a human in the person of Jesus of Nazareth (Jn. 1:1-14)? Did the Lord Jesus die for our sins on the Cross and rise physically from the dead (1 Cor. 15:3-7)? And will we meet Jesus after we die to be judged by him at the end of time (Rev. 20:11-15)?

Liberal Christians answer "no" to all these questions. They believe that Jesus was just a man and not God. However, conservative Christians throughout church history, like St. Augustine, Martin Luther, and John Calvin, have answered these three questions with an unwavering and resounding "YES." Belief in the divinity of Jesus, his sacrificial

THE BIBLE & ANCIENT SCIENCE

death and physical resurrection, and his judgment of men and women at the end of time is what makes a person a conservative Christian, like me.

We can now answer the question in the title of this hermeneutical principle: Does Conservative Christianity Require Scientific Concordism? My answer is "no." Christian faith is about Jesus, and Jesus alone. To use a well-known aphorism, Christianity is about the Rock of Ages (God), not the age of the rocks.

Biblical Creation Accounts

The so-called "conservative" Christian view of biological origins for twenty centuries of church history has been the *de novo* creation of plants and animals. Genesis 1 clearly states ten times that living organisms were created "according to their/its kinds." As well, the "conservative" position on human origins during this period firmly upheld the *de novo* creation of Adam and Eve. Genesis 2 asserts that the first man was made from the dust of the earth and the first woman from his side. In fact, the quick and complete creation of living organisms including humans was embraced by St. Augustine, Martin Luther, and John Calvin.

Augustine believed that the entire world was created in an instant with living creatures first appearing in the form of a seed, somewhat similar to the reproductive seeds of preformatism (1-seed model). These seeds then developed into fully-formed plants and animals. Luther in appealing directly to Genesis 1 argues that "the world with all its creatures was created within six days, as the words read."[22] He adds, "We know from Moses that the world was not in existence before 6,000 years ago" (Christian tradition asserts that Moses was the author of Genesis 1).[23] Similarly, Calvin accepts the age of the world to be a "period of six thousand years." In criticizing the view of St. Augustine, he argues that "Moses relates that the work of creation was accomplished not in one moment, but in six days."[24]

But as we have seen, Augustine, Luther, and Calvin had an ancient understanding of astronomy. They all believed in the firmament. So, we should not be surprised that these eminent Christian scholars also embraced an ancient view of biological origins—the *de novo* creation of plants and animals. And since they are three of the most towering figures in church history, it is understandable that their view of the origin of living organisms, including humans, became the "conservative" Christian position. This explains why in the United States 61% of Roman Catholics and 87% of Evangelical (born-again) Protestants are young earth creationists, believing that Genesis 1 is "literally true, meaning that it happened that way word-for-word."[25]

But a number of questions need to be asked. Will our views regarding how God created living creatures distinguish conservative Christianity from liberal Christianity? Or does rejecting the quick and complete origin of plants and animals in Genesis 1 and Genesis 2 make you a liberal Christian? And even more challenging, if you accept human evolution and reject that the first human male was made from the dust of the ground and the first human female from his side as described in Genesis 2:7 and 21-22, then does this mean that you can no longer call yourself a conservative Christian?

For me, *how* humans were created is incidental to the inerrant spiritual truth of *who* created men and women—the God of Christianity. More specifically, I reject the so-called "conservative" Christian view that living organisms were created *de novo*, because this is an ancient understanding of origins. In the same way that I do not accept the ancient astronomy of St. Augustine, Martin Luther, and John Calvin, I do not embrace their ancient biology regarding the origin of plants, animals, and humans. I also reject their concordist interpretations of Scripture. As we have seen throughout this book, scientific concordism always fails.

Instead, I believe that the purpose of the accounts of origins in the Bible is to reveal inerrant messages of faith. The opening chapters of Genesis reveal that God created plants and animals. In particular, the Lord made humans, we were created in the Image of God, and all of us have sinned against him and against other humans. I am certain that followers of Jesus will agree that these are some of the most important foundational beliefs of their conservative Christian faith.

HERMENEUTICAL PRINCIPLE 20

Literary Criticism

Literary criticism analyzes the various characteristics found in literature, including the Word of God. First and foremost, it focusses on the genre of a written work. As we noted in Hermeneutical Principle 2, the Bible has many types of literature such as parables, poetry, allegories, stories, historical accounts, etc. Determining the genre *dictates* interpretation. Literary criticism also examines the themes and plots in a text that together communicate the main message intended by an author. This analysis investigates the use of language, stylistic techniques, and special terminology. And literary criticism identifies the structures and arrangements of words and sentences within a written text.

In this hermeneutical principle, we will analyze the literary features of the two main accounts of origins in the Bible—the six days of creation and day of rest in Genesis 1:1 to 2:3, and the creation of Adam and Eve and the events in the Garden of Eden in Genesis 2:4 to 3:24.

Ancient Poetry & Genesis 1

The popular definition of the word "poetry" refers to literature that has figurative language and fanciful ideas. As a result, many Christians assume that statements about nature in poetic passages in Scripture do not align with physical reality. They contend that these passages are not to be understood literally and can be written-off as mere figures of speech. In many Sunday schools, this is commonly known as the "poetic language argument."

For example, the Book of Psalms has highly structured poetic hymns and prayers. As noted previously, Psalm 148:3-4 records, "Praise him [the Lord], sun and moon; praise him, all you shining stars. Praise him, you highest heavens and you waters above the skies." Some modern Christians might be tempted to write-off the reference to "waters above the skies" (i.e., the heavenly sea) as merely "poetic" and "figurative." But to be consistent, they should also dismiss the physical reality of the sun, moon, and stars. I doubt any Christian would do so.

The correct and most basic definition of the term "poetry" simply means "structured writing." As Figure 20-1 reveals, the Genesis 1 creation account is structured on a pair of parallel panels. This ancient poetic framework is arranged to highlight God's creative acts in dealing with the pre-creative state described in Genesis 1:2. The earth in this verse is described using the Hebrew adjectives *tōhû* and *bōhû*, meaning "formless" and "empty," respectively. These rhyming words would have caught the attention of the Hebrew readers and listeners. In the first three days of creation, the Creator responds to the formlessness by set-

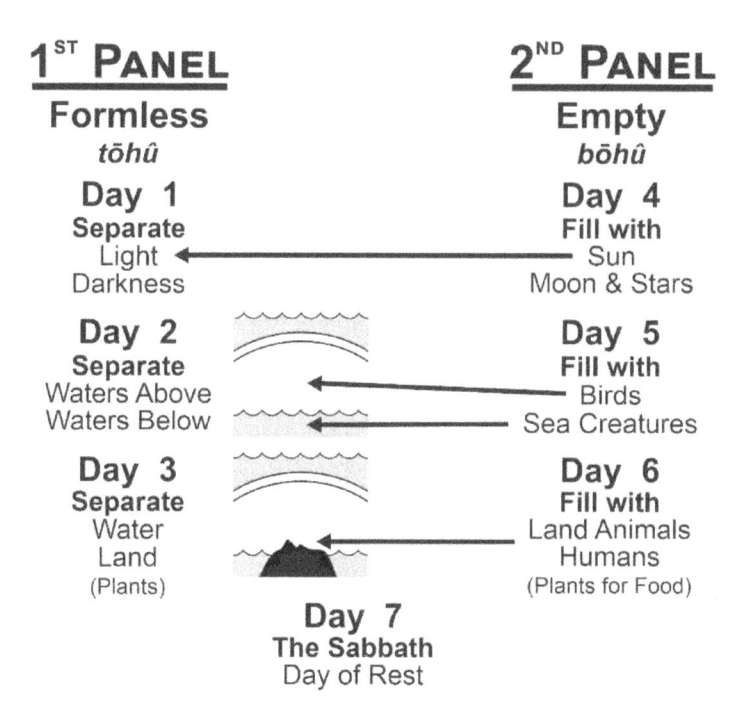

1ˢᵀ PANEL
Formless
tōhû

Day 1
Separate
Light
Darkness

Day 2
Separate
Waters Above
Waters Below

Day 3
Separate
Water
Land
(Plants)

2ᴺᴰ PANEL
Empty
bōhû

Day 4
Fill with
Sun
Moon & Stars

Day 5
Fill with
Birds
Sea Creatures

Day 6
Fill with
Land Animals
Humans
(Plants for Food)

Day 7
The Sabbath
Day of Rest

Figure 20-1. Parallel Panels of Genesis 1

ting up the boundaries of the universe. During the last three creation days, he resolves the emptiness through filling the world with heavenly bodies and living creatures.[1]

Obvious parallels exist between the two panels in Genesis 1. On the first day of creation, God creates light in alignment with the fourth day's creation of the sun, moon, and stars. Use of the firmament to separate the heavenly "waters above" from the earthly "waters below" on the second creation day forms an air space for flying creatures and a body of water for marine creatures, which are both made on the fifth day. And during the third creation day, the Creator commands dry land to appear out of the sea and then the earth to produce plants and fruit-bearing trees. These creative events match up with the sixth day of creation and the origin of land animals and humans, as well as meeting their need for food.

Another poetic feature appears in the descriptions of each day of creation in Genesis 1. There is a structured and repetitive formula that basically follows the pattern:

 (1) Introduction: "And God said, '. . .'"
 (2) Divine command: "Let there be . . ." or "Let the . . ."
 (3) Statement of completion: "And it was so."
 (4) Divine evaluation: "And God saw it was good."[2]
 (5) Time reference: "And there was evening, and there was morning."

In addition, Genesis 1 uses repeatedly the stylistic number 7 and its multiples. This number in the ancient Near East was considered special and carried a sense of completion, fulfillment, and perfection.[3] For example, the Hebrew divine name *'Ĕlōhîm* that is translated as "God" appears 35 times (7 X 5). The total number of words in this account of creation is 490 (7 X 70). The word "earth" is found 21 times (7 X 3) and "heaven/firmament" 21 times (7 X 3). Genesis 1:1 has 7 Hebrew words and Genesis 1:2 has 14 words, making a total of 21 (7 X 3). The 7th day of the week for the people of Israel was the Sabbath. It was a holy day and the day of rest. The 7th day is mentioned 3 times in Genesis 2:2-3 and each time in a sentence with 7 words, adding up to 21 words (7 X 3).

LITERARY CRITICISM

Most prominently, Genesis 1 occurs over 7 days and emphasizes the 7th day, the day God rested from this creative activity.

If we recognize and respect the ancient parallel panels in Genesis 1, then we can easily address the so-called "contradiction" often launched at Christians by anti-religious people. These critics argue that Scripture cannot be true because it is not possible for light to exist on the first day of creation three days before the creation of the sun on the fourth day. However, the reason light is created on the first day is because the first panel in Genesis 1 deals with the Creator setting up the boundaries of the world. In this case, the separation of light from darkness. The filling of the world with objects like the sun only begins in the second panel on the fourth day of creation. This arrangement of God's creative acts reflects the ancient poetic framework of parallel panels.

And I might add, are anti-religious critics of Scripture and Christianity so naïve and short-sighted that they truly think ancient people did not know that there was a connection between the sun and the appearance of light? This fact of nature was well within the scope of cognitive competence of the biblical writer of Genesis 1 and his generation. Therefore, the alleged "contradiction" of light being created before the sun is solid evidence of *poetic license* or *literary freedom* on the part of the biblical writer. It clearly indicates that this creation account is not a strict literal record of actual events in origins. In other words, the Holy Spirit-inspired author of Genesis 1 never intended to offer a list of the divine creative acts in a chronological sequence.

A comment also needs to be made regarding the word "day" in Genesis 1. The Hebrew noun *yōm* can refer to a 24-hour day and also to a period of time including many days. About ninety-five percent of the times the singular form of this word appears in the Old Testament, the context indicates that it refers to a regular day. Progressive creationists (or day-age creationists) claim the six days of creation represent six periods of time that are millions of years long. However, when the Hebrew word *yōm* appears in the Old Testament with a number, as it does in Genesis 1, it refers to a 24-hour day. In addition, Genesis 1 defines each creation day with a time reference: "And there was evening, and there was morning—the first [second, third, etc.] day."

The interpretation of the days of Genesis 1 as 24-hour days is further supported by the fact that Genesis 1 is also structured on the Hebrew work week and Sabbath. By having the Creator make the world in the first six days and then rest on the seventh day, the biblical author is affirming the Fourth Commandment that we must take a day to rest and honor God. Obviously, the Sabbath day is a regular 24-hour day and not a period that is millions of years in length. In addition, if progressive creationists want to claim that Genesis 1 is a record of God's creative acts over great periods of time, then plants created during the third "day" or age will not have sunlight to survive because the sun only appears millions of years later in the fourth "day"/age. Using the Bible as a book of science always fails. In sum, the days of Genesis 1 are 24-hour days.

Young earth creationists are quick to use the Sabbath Commandment to defend their literal and scientific concordist interpretation of Genesis 1. In Exodus 20:8 and 11, God orders the people of Israel, "Remember the Sabbath day by keeping it holy . . . For in six days the Lord made the heavens and the earth, the sea, and all that is in them, but he rested on the seventh day. Therefore, the Lord blessed the Sabbath day and made it holy." To be sure, this seems to be a powerful argument for young earth creation, because the Fourth Commandment appeals to a literal and concordist reading of the Genesis 1 creation account in order to support the Sabbath and the practice of taking a day of rest.

However, there is a subtle and fatal problem with this argument. It fails to identify the ancient science in the Sabbath Commandment. When Exodus 20:8 and 11 point back to Genesis 1 and the creation of "the heavens and the earth, the sea," it was referring to the *de novo* creation of a 3-tier universe. Similarly, the creation of "all that is in them" includes the *de novo* creation of plants, animals, and humans. But as we have seen, quick and complete creation is an ancient understanding of origins and it does not align with physical reality. Therefore, the Fourth Commandment in Exodus 20 and the appeal to Genesis 1 must be understood in

Exodus 20:8 & 11

→ **MESSAGE**
Spiritual Truths
God is Creator of Everything
Remember the Sabbath
Keep it Holy

→ **INCIDENT**
Ancient Origins
De Novo Creation of the
Heavens, Earth, Sea &
Everything in Them

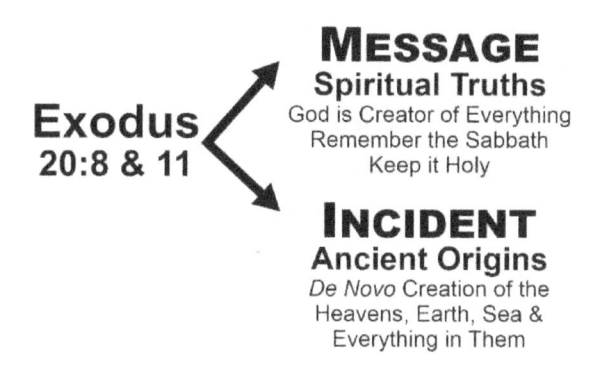

Figure 20-2. Exodus 20 & the Message-Incident Principle

light of the Message-Incident Principle, as depicted in Figure 20-2. The *de novo* creation of the cosmos and living organisms is incidental, and it must be separated from, and not conflated with, the inerrant spiritual truths that God is the Creator and we must honor the Sabbath.

To conclude, the Genesis 1 account of creation features ancient poetry. This chapter is framed on a pair of parallel panels with each creation day following a structured and repetitive formulation. But as everyone knows, actual events in the past do not unfold in such a patterned and artificial manner. Poetic license and stylistic numbers also point away from this creation account being a scientific record of the origin of the universe and life. To be sure, the author of Genesis 1 certainly believed that God created the world *de novo*, but his literary freedom allowed him to place these creative acts in a poetic order that ultimately affirms the Sabbath Commandment.

Parable-Like Story & Genesis 2-3

Throughout the history of the church, most Christians have firmly believed that the creation of Adam and Eve and the events described in the Garden of Eden in Genesis 2-3 actually happened during the past. In particular, Genesis 2 is often seen as an elaboration of the brief description of human origins on the sixth day of creation in Genesis 1. Clearly, this traditional interpretation is deeply rooted in scientific concordism. According to young earth creation and progressive creation, these open-

THE BIBLE & ANCIENT SCIENCE

ing chapters in Scripture are biblical proof against human evolution. However, there is a critical question that needs to be answered before accepting this anti-evolutionist interpretation. What is the literary genre of Genesis 2-3? And there is a related question. Are there literary features in these two chapters that give us clues as to their correct interpretation?

To begin answering these questions, let me propose a thought experiment. Assume for a moment that the Bible did not include the account of Adam and Eve in the Garden of Eden. And let's also presuppose that you have just discovered an ancient document with an account of origins and the following features:

- *de novo* creation of a garden paradise without suffering and death
- *de novo* creation of one man and one woman
- a man made from the earth and a woman from his side
- word play with a man named "earthling" and a woman named "life" or "mother of life"[4]
- all people on earth descend from this original pair of humans
- *de novo* creation of birds and land animals made from the earth
- a mystical tree with fruit that imparts knowledge of good and evil
- a mystical tree with fruit that imparts eternal life
- a fast-talking snake that tempts a woman to disobey the one and only command of God
- suffering and death enter the world for the first time because God judges one man and one woman for disobeying his one and only command
- mystical creatures with wings, the body of a lion, and a human head (these are the "cherubim" and they are like the Sphinx in Egypt)
- a flaming sword flashing back and forth to guard the way to the mystical tree with fruit that imparts eternal life

After reading this account of origins, would you immediately conclude that it is a historical record of actual events from the past? Would you then go to the history department at a university to claim that you have discovered a document that needs to be added to textbooks on the history of humanity? Or would you say that all these features indicate this account is a story with spiritual lessons that includes ancient science and allegorical characteristics? If we can suspend our belief in the existence of Genesis 2-3, then I think that most people would say your newly discovered ancient account of origins is a made-up story somewhat like a parable, but not real history.

Of course, you know that these features above appear in Genesis 2-3 with the creation of Adam and Eve and the events in the Garden of Eden. I believe that if we look beyond traditional literal and concordist interpretations of these chapters, and focus instead on these literary characteristics, they become clues in determining the genre of this biblical account. For me, Genesis 2-3 is a Holy Spirit-inspired story with features that are to some extent similar to the parables of Jesus. Most importantly, these chapters offer life-changing inerrant spiritual truths. But in being a story with numerous allegorical characteristics, this indicates that the events in Genesis 2-3 never actually happened in the past.

To be sure, my view shocks most Christians. And I apologize if this offends you. But I'll ask for your patience, please. If you will recall Hermeneutical Principle 1, I presented evidence that literal passages in the Bible are not more important or holier than non-literal passages. Jesus' use of stories, parables, and figurative language is proof that God has employed non-literal forms of literature in Scripture. In fact, about one third of the Lord's teachings are parables. These are stories in which the events that are mentioned never really happened (e.g., parable of the Good Samaritan). With this being the case, it leads us to consider the possibility that the Holy Spirit inspired some human authors of the Bible to employ made-up stories in order to reveal inerrant spiritual truths. Let me offer two examples.

The Book of Job is the famed biblical book that deals with the problem of suffering. Most Christians assume that there really was a man named "Job" and that events recorded in this book really happened. However, the events described in the opening two chapters raise some serious questions. First, are we to believe that Satan can just appear in heaven and bait God to prove a point? In particular, would the Lord allow Satan to murder Job's ten children and his servants in order to test Satan's assumption that Job would curse God for these events? Does this sound like something God would permit? Satan is then allowed to inflict Job with a dreadful disease with the same intention of inciting him to curse the Lord. Would Jesus let Satan do this?

Or is the Book of Job a masterfully crafted Holy Spirit-inspired story that was made up to reveal spiritual truths about the challenging issue of human suffering? From chapters 3 to 37, Job and his three friends attempt to justify his suffering by using various misguided assumptions and arguments. But in chapters 38 to 41, God speaks directly to Job. Notably, he does not give Job an explicit reason for his suffering. Instead, the Creator simply points to his marvellous and intelligently designed world. The creation humbles Job and he repents, "Surely I spoke of things I did not understand, things too wonderful for me to know" (Job 42:3). The powerful revelation in nature discloses that God is in complete control of the world, including our suffering, even though we may not fully comprehend why at times we experience dreadful pain and heartache (I am writing this sentence during the Covid-19 pandemic). Instead of the horrid events at the beginning of the Book of Job having actually happened, they are part of the literary genre of a story, a made-up account that is a brilliantly effective instrument in delivering these spiritual truths.

The Book of Jonah is another well-known book in Scripture. God called Jonah to preach in the city of Nineveh, but instead he sailed away on a ship to the ends of the known world. During the voyage a violent storm arose. Being aware that Jonah was disobeying God, his shipmates threw him overboard in order to stop the storm. Then "the Lord provided a huge fish to swallow Jonah, and Jonah was in the belly of the fish three days and three nights" (Jon. 1:17). How could Jonah have survived that long in the belly of a fish without air and engulfed in stomach acid?

Nearly every Christian knows about Jonah's three days in a fish. This demonstrates the effectiveness of stories that feature astonishing events. They are easy to remember, including their main messages. This is similar to Jesus using hyperbole in the Sermon on the Mount (e.g., plucking out our eyes when we lust). It is also the technique employed in

LITERARY CRITICISM

the Book of Job, when Satan murders Job's children and servants, and inflicts Job with horrid physical conditions. But more importantly, it is the spiritual message delivered by the incredulous account of Jonah being swallowed by a fish. Jonah's story is the story of many people. And I know this personally. We run from God and his calling on our life, yet he graciously saves us from our foolish disobedience and even offers us other opportunities to follow his will.

Great stories like those in the Books of Job and Jonah are archetypal. The Greek noun *archē* means "beginning," "origin," and "the first cause;" and *tupos* refers to "type," "pattern," and "model." An archetype is a first or original person, thing, or event that represents other members of a group. The stories of Job and Jonah typify different aspects of the human spiritual condition. They resonate deeply with our life experiences, and we personally identify with the people and events described in these stories. Has any Christian who has endured suffering not seen him or herself as Job? Or, who has not disobeyed God like Jonah and ran away from the Lord's will and calling for their life?

Similar to Job and Jonah, Adam and Eve are archetypes. The first man and first woman in the Bible embody a central aspect of our spiritual essence—we are all sinners. The Garden of Eden story, with its allegorical elements, is a perfect picture of human rebellion against the Creator. God creates a garden paradise for humans. He gives Adam just one commandment: do not eat fruit from the mystical tree of the knowledge of good and evil. But tempted by a talking snake, Eve eats the fruit of this tree and Adam does as well. God then calls them both to accountability. They attempt to rationalize their disobedience. The woman blames the snake. The man blames the woman, and he even subtly blames the Lord for putting her in the garden with him! They are then banished from the garden.

Remarkably, in just two short biblical chapters, the story of Adam and Eve deals with many central spiritual truths of the Christian faith—divine command, temptation, disobedience, accountability, rationalization, and alienation. Like the parables of Jesus, the Garden of Eden story powerfully impacts us by revealing who we really are, even though the events in Genesis 2-3 never really happened. The Lord gives us the gift of life, yet we disobey his commands. And then we try to rationalize our sins by playing "the blame game." If we cannot see ourselves as Adam and Eve, then we need to re-read Genesis 2-3 until we recognize that they represent each and every one of us.

The Interpretive Key for Creation Accounts

For more than forty years, I immersed myself in hundreds of creation accounts from throughout the ancient world, especially those of ancient Near Eastern countries surrounding ancient Israel.[5] I am embarrassed to say that it is only in the last couple of years that I have come to appreciate the key to their interpretation. And here is what I finally understood: *Ancient authors in their accounts of origins featured parable-like stories that they made up.*

To be sure, *de novo* creation is the view of origins in most of these ancient accounts. The God or gods made the universe and living organisms quickly and completely into their mature forms. This was the science-of-the-day in the ancient world. But more importantly, these creation accounts focus on made-up stories that are similar to parables, and these stories deliver the most cherished religious and philosophical beliefs of the community or civilization. Ancient authors were skillful storytellers and the messages they told through their well-crafted stories were critical in holding their people together in order to function as a society. For example, the ancient Hebrew community was bound together by the powerful belief in a Creator who was a holy God.

Of course, most of us as Christians have failed to identify the story component in the opening chapters of the Bible, especially the account of Adam and Eve in Genesis 2-3. And worse, we have turned the made-up and parable-like story of the Garden of Eden into a scientific and historical account of events that really happened in the past. This is no different than taking the story of Jack and the Bean Stalk and claiming it is part of human history. I doubt anyone would do this today. As twenty-first century Christians, we tend to overlook the Holy Spirit-inspired stories in Genesis 1-3. Why is this the case?

LITERARY CRITICISM

I think there are three main reasons. First, most conservative Christians embrace scientific concordism because this has been a longstanding tradition throughout church history. As most of you know, this hermeneutical approach is taught within churches and Sunday schools. Second, we are a scientific society and we very much value scientific facts and discoveries. We then assume that since the Bible is the Word of God, it must be in alignment with modern science in some sort of way. Third, and closely related to the previous point of science dominating the modern mindset, stories are devalued in our culture. They are often viewed to be merely for our entertainment, like novels and movies. Others see stories as nothing but cute tales for children with a moral lesson. In other words, made-up stories with parable-like features are not taken seriously.

And I might add a fourth reason, which is personal. Scientists who study origins today never include made-up stories in their scientific publications. In my dozen or so scientific papers on dental development and evolution, my co-authors and I have never incorporated stories with philosophical or religious messages.[6] Science is limited to exploring the physical world, and the physical world only. Therefore, as a modern culture we do not have examples of modern scientific accounts with made-up and parable-like stories revealing our beliefs and worldview. This is a literary genre that simply does not appear in our culture today. It is understandable, therefore, why modern Christians, like me for so many years, fail to identify the stories in the opening chapters of the Bible. Most of us are not aware these chapters are a distinctive *ancient literary genre*.

To summarize, the Holy Spirit-inspired authors of Genesis 1-3 certainly believed in the *de novo* creation of the universe and living organisms, including a first man and a first woman. Yet as masterful storytellers, they had the literary freedom to make up an account of events in order to reveal as effectively as possible inerrant spiritual truths. Therefore, our goal as twenty-first century readers of the Word of God is to separate the parable-like stories and the ancient science of origins in the biblical creation accounts from the life-changing messages of faith, and not to conflate these three components together.

Adam & the Apostle Paul

Of course, I suspect most of you are asking, "What about the apostle Paul? Did he believe that Adam was a real person from the past? Or did he understand that Genesis 2-3 was a made-up story and not historical?" As we have noted previously, Paul accepted the 3-tier universe as seen in Philippians 2:10—"that at the name of Jesus every knee should bow, [1] in heaven and [2] on earth and [3] down in the underworld." Since Paul embraced ancient geography and ancient astronomy, it is only consistent that he also accepted ancient biology and the *de novo* creation of humans. This was the science-of-the-day.

Now, there is another important aspect that we need to appreciate regarding Paul's views in order to understand his interpretation of Genesis 2-3. During the intertestamental period (from the completion of the Old Testament around 400 BC to the New Testament in the 1st century AD), Jewish literature reveals that biblical interpretation firmly accepted scientific concordism and the historicity of Adam. For example, the Book of Tobit (about 200 BC) states, "You [God] made Adam, and for him you made his wife Eve as a helper and support. From the two of them the human race has sprung."[7] A concordist hermeneutic also appears in the Book of Sirach (after 200 BC). "The Lord created human beings out of earth, and makes them return to it again . . . From a woman sin had its beginning, and because of her we all die."[8] Wisdom of Solomon (about 150 BC) records, "For God created us for incorruption, and made us in the image of his own eternity, but through the devil's envy, death entered the world."[9] And a contemporary of Paul, the Jewish historian Flavius Josephus (37-100 AD), wrote a history of Israel in his *Antiquities of the Jews*. In the preface of his book, he clearly indicates that his authorial intention is to write "our history" and begins with basically a paraphrase of the opening chapters of Genesis and all humanity descending from Adam.

Therefore, concordism was the hermeneutics-of-the-day in the Jewish community just prior to and around the time that Christianity arose.[10] As a consequence, Paul undoubtedly believed that Adam was a historical person and that the events in Genesis 2-3 really happened. However, it must be emphasized that Paul's belief in the reality of Adam and the events in the Garden of Eden does not necessarily mean they are historical. Remember he accepted the 3-tier universe. Does Paul's belief indicate that the structure of the world actually has three levels—the heaven, the earth, and the underworld? I doubt any Christian today would use this argument. This logic also applies to Adam and the events in Genesis 2-3. Paul's belief in their historicity does not mean they really existed.

But more importantly, Paul uses Adam as an archetype to deliver spiritual truths. In Romans 5:14, he states that Adam "is a pattern of the one to come [i.e., Jesus]." The Greek noun translated as "pattern" is *tupos* and also means "type."[11] Paul further explains in 1 Corinthians 15:45, 47, and 49. "So it is written, 'The first man Adam became a living being'; the last Adam [Jesus], a life-giving spirit . . . The first man was of the dust of the earth; the second man is of heaven . . . And just as we have borne the image of the earthly man, so shall we bear the image of the heavenly man." In other words, the "first man Adam" is the archetype of the human sinner, while the "last Adam" Jesus is the archetype whose likeness men and women, by grace, will become.

Excursus
Jesus, Adam & Genesis 1-2

In Hermeneutical Principle 1, I introduced one of the most challenging passages in the Bible to interpret. Matthew 19:1-12 records an encounter between Jesus and the Pharisees regarding divorce. They asked him in verse 3, "Is it lawful for a man to divorce his wife for any and every reason?" The Lord responds in verses 4-5 by appealing to the creation of humans in Genesis 1 and 2. "Haven't you read," he [Jesus] replied, "that at the beginning the Creator 'made them male and female' [Gen. 1:27], and said, 'For this reason a man will leave his father and mother and be united to his wife, and the two will become one flesh [Gen. 2:24]?'"

On the surface, it does seem like Jesus is affirming historical reality of the *de novo* creation of humans in Genesis 1, and in particular the existence of Adam and Eve in Genesis 2. However, the authorial intention of Jesus in this passage is not to affirm whether or not people were actually made quickly and completely, or whether or not Adam and Eve really existed. This encounter with the Pharisees was not a debate about origins. Instead, the issue that was being discussed was divorce.

To defend his position on divorce, Jesus employs two arguments. First, he accommodates by employing the ancient science of the *de novo* creation of "male and female" in Genesis 1:27. In this way, Jesus emphasizes the inerrant spiritual truth that God created human beings, and since God is the Creator of men and women, he is also the Lord of our life. Consequently, we are accountable to him regarding how we live, including the relationship between a husband and a wife. Second, Jesus uses the story of Adam and Eve as an archetype to affirm God's plan that a healthy marriage is a lifetime commitment between one man and one woman. As the wonderful metaphor in Genesis 2:24 states, through marriage a husband and wife "become one flesh."

Figure 20-3 applies the Message-Incident Principle to Matthew 19:1-12. This passage is not a scientific revelation concerning how God actually made humans. Again, it is not a debate about origins. Instead,

Matthew 19:1-12
→ **MESSAGE**
Spiritual Truths
God Created Men & Women
Instituted Marriage &
Opposes Divorce

→ **INCIDENT**
Ancient Origins
De Novo Creation of
Adam & Eve

Figure 20-3. Matthew 19 & the Message-Incident Principle

Jesus was responding to frivolous excuses justifying divorce. By appealing to Genesis 1:27 and 2:24, he argues that divorce was never God's intention for a husband and wife. And concluding his argument in Matthew 19:6, the Lord commands, "What God has joined together, let no one separate." Indeed, this is an inerrant spiritual truth that needs to be heard and obeyed in our culture today.

HERMENEUTICAL PRINCIPLE 21

Source Criticism

Source criticism analyzes literature in order to determine if it includes passages from earlier documents. The notion that biblical authors used sources is explicitly mentioned in Scripture. For example, Numbers 21:14-15 states, "That is why the Book of the Wars of the Lord says, '. . .'" This book no longer exists (or has yet to be found) and presumably is a military history including a collection of victory songs. Likewise in 2 Samuel 1:17-18, "David took up this lament concerning Saul and his son Jonathan, and he ordered that the people of Judah be taught this lament of the bow (It is written in the Book of Jashar): '. . .'" Verses 19-27 then reproduce the song from this ancient document.

The use of sources is also mentioned in the introduction to the Gospel of Luke. In Luke 1:1-4, the inspired author explains the purpose of his book and the method he employed to write it.

> Many have undertaken to draw up an account of the things that have been fulfilled among us [the life of Jesus], just as they were handed down to us by those who from the first were eyewitnesses and servants of the word. With this in mind, since I myself have carefully investigated everything from the beginning, I too decided to write an orderly account for you, most excellent Theophilus, so that you may know the certainty of things you have been taught.

Like a modern historian, Luke utilized various sources that "were handed down" to him and "carefully investigated everything from the beginning" so that he could "write an orderly account." These sources likely included both oral testimonies and written reports.

It must be emphasized that the use of sources by biblical authors does not in any way weaken or undermine the belief that Scripture is the Holy Spirit-inspired Word of God. Jesus himself assured his disciples in John 14:26 that "the Holy Spirit, whom the Father will send in my name, will teach you all things and will *remind* you of everything I have said to you" (my italics). Therefore, the Lord guaranteed that his teaching would be preserved by his disciples. As the first eyewitnesses of the ministry of Jesus, they handed down accounts of his life and teaching to others like the gospel writer Luke. Similarly, 2 Timothy 3:16 states that "all Scripture is God-breathed" and 2 Peter 1:21 asserts "prophets, though human, spoke from God as they were carried along by the Holy Spirit." In this way, the process of biblical revelation, including the use of sources, was ultimately inspired by God.

Biblical Creation Accounts

There is convincing evidence that indicates the creation accounts in Genesis 1 and Genesis 2 come from two different original sources. You might have suspected that this is the case in the previous hermeneutical principle with the former chapter featuring the ancient poetic framework of parallel panels, and the latter characterized by a parable-like story. Differences in the order of creative events, writing style, terminology, scene-setting, and theological emphasis between these two chapters point toward two distinct Holy Spirit-inspired writers.

These two authors are usually referred to as the "Priestly (P) writer" of Genesis 1 and the "Jahwist (J) writer" of Genesis 2. The J source is often dated around 1000 BC, when a stable society in Israel emerged during the reign of King David. The date of the P source is about 500 BC, after the Jews returned from the exile in Babylon (597–538 BC). Afterward, an editor who is often termed the "redactor" combined the Priestly and Jahwist sources into what has become the opening chapters

THE BIBLE & ANCIENT SCIENCE

of our Bible. Let's examine some biblical evidence that points to these two sources.[1]

The order of God's acts of creation in Genesis 1 and Genesis 2 do not align, which suggests these are two different accounts of origins. This conflict is particularly obvious with the creation of humans. In Genesis 1, man and woman are created *after* vegetation, fruit trees, birds, and land animals. But in Genesis 2, the creation of a man is *before* vegetation, fruit trees, birds, and land animals.

Genesis 1	**Genesis 2**
vegetation & fruit trees	man
birds	vegetation & fruit trees
land animals	land animals & birds
man & woman	woman

The origin of birds is one of the most glaring problems with the traditional and scientific concordist interpretation of Genesis 2 being an elaboration of the events on the sixth creation day in Genesis 1. According to Genesis 1:21, God created "every [*kol*] winged bird according to its kind" on the fifth day of creation, one entire day *before* the creation of humans. However, Genesis 2:19 asserts that the Lord God formed "all [*kol*] the birds in the sky" *after* the man was made in Genesis 2:7. In both verses, the identical Hebrew adjective *kol* is used to indicate the creation of each type of bird ever made by the Creator.

There is also a conflict in the creative order of humans and land animals. On the sixth day of creation, Genesis 1:24 places the origin of "the livestock, the creatures that move along the ground, and the wild animals" *before* the creation of male and female humans, described in Genesis 1:27. But Genesis 2:19 puts the formation of "all the beasts of the field"[2] *in between* the fashioning of Adam from the ground in Genesis 2:7 and Eve from his side in Genesis 2:22.

The origin of vegetation and fruit trees is another significant contradiction between the Genesis 1 and Genesis 2 creation accounts. On the third day in Genesis 1:11, God made plants and trees, including trees "that bear fruit," three days *before* the creation of humans. However, Genesis 2:5 states "no shrub had yet appeared on the earth and no plant had yet sprung up." The Lord God then creates Adam in Genesis 2:7, and *afterwards* plants a garden with fruit trees in Genesis 2:8.

These conflicts in the order of God's creative acts certainly challenge the traditional and scientific concordist interpretation that Genesis 2 is an expanded description of the events on the sixth day of creation in Genesis 1. However, this problem quickly disappears if we recognize that the Holy Spirit inspired the editor or redactor of the Book of Genesis to use two different, yet theologically (spiritually) complementary, creation accounts in the process of biblical revelation.

The contradictory order of creative events in Genesis 1 and Genesis 2 is a serious challenge to scientific concordism, in that at least one version of the creative events is obviously incorrect. For example, either birds were created before humans (Gen. 1), or they were made after the man and before the woman (Gen. 2). Both accounts cannot be true. From my perspective, instead of weakening and undermining Scripture, these conflicts are evidence *within Scripture itself* that God never intended to reveal how he actually created living organisms. To be more precise, I believed the Holy Spirit allowed these contradictions to point us away from using the Bible like a book of science. Thus, the order of God's *de novo* creative acts in these two accounts of creation is ultimately incidental and not relevant to our faith.

There are also significant differences in literary style between Genesis 1 and Genesis 2. The Priestly author employs an ancient poetic format. As we discovered in the previous hermeneutical principle, the six days of creation in Genesis 1 are structured on a pair of parallel panels (Fig. 20-1, p. 163), and each of these days features a repetitive formula. Moreover, the P writer does not employ allegorical features such as mystical trees and mystical animals.

In contrast, the Jahwist author of Genesis 2 uses the genre of a story with a free-flowing style (note Genesis 2-4 is one literary unit). This creation account is somewhat like a parable and includes allegorical characteristics. There are two mystical trees, the fruit of one imparts knowledge of good and evil and that of the other eternal life, an archetypal man named "earthling" and a archetypal woman called "mother of life" (see below for the explanation of this word play), and mystical creatures including a fast-talking snake and cherubim, a composite creature like the Sphinx.

The Priestly and Jahwist writers also use distinctive terminology in their account of creation. In Genesis 1, the Hebrew divine name *'Elōhîm*, which is translated as "God," appears thirty-five times (7 X 5), reflecting an emphasis on the stylistic number 7. The phrase "according to their/its kind/s" is found ten times. The P author projects an optimistic vision with his selection of terms. Two times he mentions that God "blessed" creatures and encouraged them, "Be fruitful and increase in number." God also "blessed" the seventh day. And the Priestly writer underlines the unique and special character of men and women. He states that the Creator makes humans in his "likeness" once and in his "image" three times.

On the other hand, the J author in Genesis 2 uses the Hebrew divine name *'Elōhîm Yahweh* which is translated as "Lord God."[3] He also employs word play to emphasize the archetypal humans. The Hebrew word *'ādām* rendered as "man" echoes *'ădāmâ*, meaning "earth." So, *'ādām* could be translated as "earthling." In Genesis 3, the name "Eve" (*hawwāh*) sounds similar to the verb "to live" (*hāyâ*) and the adjective "living" (*hay*), since "she was the mother of all the living" (Gen. 3:20).[4] The J author's terminology sets a pessimistic tone, especially with an emphasis on the sinful character of humans. The literary unit of Genesis 2-4 uses the word "evil" four times and "curse/d" three times. And the term "sin" appears in Scripture for the first time in Genesis 4:7; "sin is crouching at your door; it desires to have you, but you must rule over it."

In addition, the setting of the scene in Genesis 1 and Genesis 2 are quite different. The Priestly author has a grand and cosmic picture of the creation. In Genesis 1, he describes the origin of the firmament, the heavenly sea, and the sun, moon, and stars. He also explains the gathering of the earthly sea to one place and the appearance of dry land. And the P writer includes the creation of all animals that inhabit land, sea, and air.

But the scene-setting in Genesis 2 is regional and limited to a specific rural location. There is no mention of the sea or the heavenly bodies. The Jahwist author makes numerous geographical references to the ancient Near East, none of which are found in Genesis 1. Three countries are mentioned: Havilah, Cush, and Assur. And the J writer refers to the headwaters of four rivers: Pishon, Gihon, Tigris, and Euphrates. The sources of the latter two rivers are well-known and set the Genesis 2 creation account somewhere in northwestern Mesopotamia.

Finally, Genesis 1 and Genesis 2 feature distinct differences in theological emphasis. The Priestly author paints a picture of a transcendent Creator who is a Cosmic King that rules over his extensive creation. In fact, this writer uses and revamps the ancient Near Eastern concept that kings were the image of God and his representatives on earth. He states that all humans are created in the Image of God, and we are therefore all kings and queens ruling the earth and representing our Creator in heaven.[5] The P author in Genesis 1 also emphasizes the power of God. His creative action is through spoken commands, including the creation of some things out of nothing, like the firmament.

The depiction of the Lord God in Genesis 2 is an immanent Creator who is in a personal relationship with the first man and first woman. He is present on earth and fashions a man from the dust of the ground. He cares for the man and makes the woman from his side to be his companion in marriage. As Genesis 2:24 asserts, "That is why a man leaves his father and mother and is united to his wife, and they become one flesh." The Jahwist author emphasizes that the divine relationship with humans requires obedience to a Holy God. He has commandments and sets limits on human behavior and freedom. We are all accountable before him, and failure to respect his commands has serious consequences.

THE BIBLE & ANCIENT SCIENCE

	GENESIS 1	**GENESIS 2**
Literary Style	Ancient Poetry Structured & repetitive No allegorical features No word play	Parable-Like Story Free-flowing Allegorical features Word play
Setting of the Scene	Cosmic & universal Refers to the sea, sun, moon & stars No geographic references	Regional & rural No references to the sea, sun, moon & stars Refers to ANE geography
Divine Name Hebrew	God *'Elōhîm*	Lord God *Yahweh 'Elōhîm*
Creative Action	Spoken commands includes *creatio ex nihilo* eg firmament	Hands-on fashioning uses earth to make man, birds & land animals
Divine Being	Transcendent & heavenly	Immanent & earthly
Relationship to Humans	Distant & kingly	Intimate & personal
Food Commands	Focus on sustenance without a prohibition	Focus on obedience with a prohibition

Figure 21-1. Differences between Genesis 1 & Genesis 2

Figure 21-1 summarizes significant differences between Genesis 1 and Genesis 2. This is evidence that these biblical chapters were composed by two different authors. Rather than viewing these two chapters as contradictory, if we accept their separate authorship, we can see that the Bible opens with a divine revelation that God is both the transcendent kingly Creator of a vast wonderful cosmos, and at the same time, the immanent personal Maker of each and every one of us on earth. The Holy Spirit's intention by inspiring the Priestly author, the Jahwist author, and the editor (redactor) of the Book of Genesis was to reveal the fullness of God's character. In this way, the different theological emphases in Genesis 1 and Genesis 2 complement and enrich each other, giving a more complete picture of the God who graciously created us.

HERMENEUTICAL PRINCIPLE 22

Biblical Inerrancy: Toward an Incarnational Approach

The greatest act of divine revelation is the Incarnation. The Latin preposition *in* means "in" and "into," and the noun *carnis* refers to "flesh." God took on human flesh in the person of Jesus to teach us inerrant spiritual truths and to demonstrate his unfathomable love through dying on the Cross for our sins. The opening chapter of the Gospel of John eloquently summarizes the Incarnation. "In the beginning was the Word [i.e., Jesus], and the Word was with God, and the Word was God. He was with God in the beginning. Through him all things were made; without him nothing was made that has been made . . . The Word became flesh and made his dwelling among us" (Jn. 1:1-3, 14).

The early church struggled with understanding the ultimate nature of Jesus, or to use the technical philosophical/theological term, the Lord's ontology. The Greek word *ontos* is the participle of the verb "to be" and it is translated as "being." At that time there were two main heresies in the church that emerged regarding the Lord's ultimate being or essence. Docetism asserted that Jesus did not really take on human flesh but only appeared to be a man. The Greek verb *dokeō* means "to seem" and "to have the appearance." Arianism claimed that Jesus was not God, but an individual who had been created with supernatural powers. This view was formulated by Arius, a fourth century priest. Eventually the church at the Council of Chalcedon (451 AD) came to the belief that Jesus was "truly God and truly man."

In light of the Incarnation, there are some remarkable parallels between Jesus who in Scripture is called the "Word" (as quoted above) and the Bible which is referred to as the "Word of God" (Mk. 7:13; Jn. 10:35; Acts 6:2; 1 Thess. 2:13). These similarities can contribute toward an incarnational understanding of biblical inerrancy. They also offer insights into the process used by the Holy Spirit to inspire biblical writers in passages referring to the natural world. Instructive parallels appear between Scripture and: (1) the Ontology of Jesus—his divine and human nature; (2) the Temporality of Jesus—his eternity outside time and historicity within time during the first century; and (3) the Pedagogy of Jesus—his teaching of inerrant spiritual truths through the use of incidental and imperfect human words.

Firstly, Jesus is ontologically both fully divine and fully human. Being God, he is the supreme inerrant and perfect Spiritual Being. As a man, the Lord certainly experienced the physical limitations and challenges of having a human body. He felt hunger and thirst, became tired and sleepy, had common aches and pains, and suffered death (e.g., Mk. 11:12; Lk. 9:58; Jn. 4:6, 9; Matt. 27:50). I suspect that most Christians would agree that whether Jesus was six-foot-two or five-foot-two, or whether his eyes were blue or brown, his height or eye color were ultimately incidental. In this way, God's greatest act of divine revelation came through a fallible and imperfect human vessel.

The Bible has similarities to Jesus in that it is both divinely inspired and written by humans.[1] As we noted in Hermeneutical Principle 10, the New Testament was composed in Koine Greek. This ancient language is an unrefined form of Greek that was used in the streets by common people. Today it is a dead language because it is not spoken by any community. Some Christians might assume that the Holy Spirit would have employed sophisticated Greek in Scripture. But God did not do so. Our primary source of knowledge about Jesus comes through this rough earthy language. Despite the limitations and imperfections of Koine Greek, the effectiveness of this incidental ancient vessel in transporting divinely inspired spiritual truths is proven by the fact that the New Testament throughout history has powerfully impacted men and women and led them into a personal relationship with the Lord.

Secondly, Christian faith asserts that Jesus is eternal and transcends time, and that he entered within time and human history. By leaving the divine realm of eternity, God accommodated and came into the creaturely boundaries of time. As a result, the Lord's life and ministry were adapted to the culture of ancient Israel in the first century AD. He had a common job as a carpenter, ate regular foods like bread and fish and drank wine, and he taught with parables by using familiar ideas such as the science-of-the-day (e.g. mustard seed assumed to be the smallest of all seeds). It is conceivable that Jesus could have come at another point in history with him revealing the same inerrant spiritual truths. For example, if he were in America today, he might have a computer job, would probably consume sodas and hamburgers, and undoubtedly modern science would be used in some of his parables. In other words, the point in history when the Lord came into the world is ultimately incidental.

Like the temporality of Jesus, the Bible both transcends time and is bound within time and history. Scripture offers timeless eternal truths that were written down during different historical periods over roughly 1500 years. The actual points in history when the Holy Spirit inspired the biblical writers are incidental to the inerrant messages of faith. That is, there is nothing inherently special about any specific point in the past. In the same way that the Lord's timeless nature rises above his first century historicity, the eternal inerrant truths in Scripture transcend the incidental ancient historical conditions during which God revealed them. Again, evidence for this fact appears in the countless lives forever changed by the Bible in every generation. The inerrant messages of faith in Scripture were not only relevant for people in the past, but also for us today, and will be for men and women in the future.

Thirdly, Jesus taught the inerrant Word of God using the imperfect words of humans. To deliver his messages of faith as effectively as possible, he accommodated to the intellectual level of the men and women of his generation. The Lord often utilized parables and non-historical stories. Notably, he employed imperfect ancient science in his teaching

ANCIENT GEOGRAPHY

Earth is flat	Matt 4:1-11
Earth has ends	Matt 12:42
Underworld exists in the heart of the earth	Matt 12:40
Gates of the underworld	Matt 16:18

ANCIENT ASTRONOMY

Sun moves across the sky	Matt 5:45
Firmament is a solid domed structure overhead	Matt 24:29
Heaven has ends	Matt 24:31
Stars dislodge from firmament & fall to earth	Matt 24:29
Divine dwelling overhead	Jn 6:62
3-Tier Universe	Matt 11:23

ANCIENT BIOLOGY

Mustard seed is smallest of all seeds	Mk 4:30-32
Plant growth is controlled only by the soil	Mk 4:26-29
Seeds die before germination	Jn 12:23-24
Humans created *de novo* (quick & complete)	Matt 19:4-5

Figure 22-1. Jesus & Ancient Science

ministry. Figure 22-1 is a summary of ancient scientific ideas used by Jesus that we have examined in this book. Had the Lord lived today, he would certainly use the marvelous discoveries of modern science, including evolutionary biology. For example, in a debate over divorce with modern day "Pharisees," he might recast Matthew 19:4-5, "At the beginning the Creator made many males and many females through an evolutionary process. And for this reason, a man will leave his parents and be united to his wife, and the two will become one flesh."

Like the pedagogy of Jesus, the Bible is the Word of God delivered through human words. The Lord's repeated use of parables and stories is powerful evidence to support the belief that the Holy Spirit can reveal inerrant spiritual truths in biblical passages referring to events that never actually happened. In addition, the presence of ancient science in the teachings of Jesus indicates that it is possible for an imperfect understanding of the natural world to convey life-changing messages of faith. The Lord's use of the science-of-the-day sets a powerful precedent. It encourages us to employ modern science in theology. For example, the

BIBLICAL INERRANCY: TOWARD AN INCARNATIONAL APPROACH

Bible Statements about Nature

INERRANT Eternal Word of God Transcends Time

INCIDENTAL Imperfect Words of Humans Within Time & History

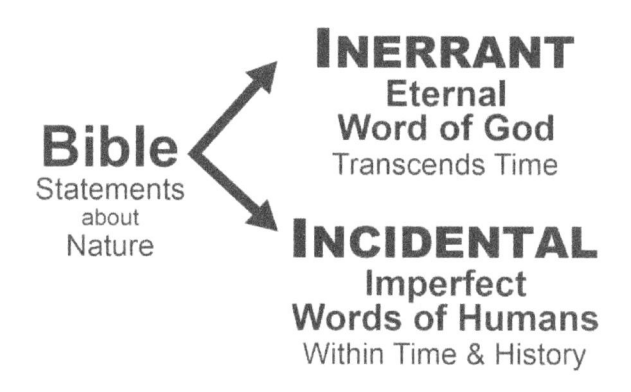

Figure 22-2. An Incarnational Approach to Biblical Inerrancy

theory of evolution could be an incidental modern vessel that transports the timeless eternal truths of the biblical accounts of creation—God created the world, humans are created in the Image of God, and all men and women are sinners.

Figure 22-2 presents an incarnational approach to biblical inerrancy for statements about nature in Scripture. The similarity to the Message-Incident Principle is evident. The message of faith is the inerrant and eternal Word of God that transcends time. The incidental imperfect words of humans are bound within time and history, like those referring to the ancient science in the Bible. In the illuminating light of the Incarnation, we can see that the inerrancy of Scripture is found in the spiritual truths, since these are absolutely true and completely free from error. This understanding of inerrancy helps us focus on the messages of faith in the Bible, and not on the less than perfect human features, such as the ancient phenomenological perspective of nature. In a manner similar to how God took on human flesh in the person of Jesus, the Holy Spirit accommodated in Scripture by using "fleshy human" ideas about the physical world as vessels to reveal himself and his love for us.

Biblical Creation Accounts

It is interesting to note that today there are interpretations of the creation accounts in Scripture that are somewhat like the early heresies regarding the ontology of Jesus. Young earth creationists and progressive creationists embrace a view of the Word of God that could be termed "biblical Docetism."[2] They contend that Genesis 1 and 2 are a perfect divine revelation that describe without any errors whatsoever how the Creator actually made the world. In other words, the science in these opening chapters is believed to be inerrant. However, these anti-evolutionists completely fail to recognize and respect the human and historically conditioned aspects in the Bible, such as the imperfect ancient understanding of origins.

There is also a modern form of "biblical Arianism." From this perspective, Scripture is seen as having no divinely inspired elements. Not only do Genesis 1 and 2 present an erroneous ancient science of the origin of the universe and life, but they also make false and mistaken statements about spiritual reality and the Creator. The Bible is viewed as merely a book of human speculations and superstitions about God. At best, it is inspiring literature, like Shakespeare and other great literary works. But it is definitely not inspired by the Holy Spirit. This assessment of the Bible characterizes liberal Christianity and its lack of respect for the Word of God.

An incarnational approach toward inerrancy and the accounts of creation in Scripture stands between "biblical Docetism" and "biblical Arianism." The Lord Jesus enlightens us and leads to a balanced understanding that acknowledges the Bible is both divinely inspired and humanly authored. To recast G.E. Ladd's aphorism presented in Hermeneutical Principle 14—the accounts of origins in Genesis 1-3 are the inerrant and eternal Word of God written in the incidental and imperfect words of men and women in history.

CONCLUSION

Beyond Scientific Concordism

I hope that you have enjoyed reading this book as much as I have enjoyed writing it. I never get tired of studying the Word of God. From an academic and analytical perspective, I am forever struck by the wisdom and foresight of the Holy Spirit when he inspired the writers during the process of biblical revelation. It makes perfect sense to me why God accommodated to the intellectual level of ancient people. This was the most effective way for the Lord to reveal himself and his love for everyone. From a personal and devotional perspective, the Bible is indeed "living and active" (Heb. 4:12). My day begins in prayer through reading Scripture. I always hear the "voice" of God arise from its pages, comforting and encouraging me, as well as challenging and cautioning me.

Let me repeat and emphasize that the hermeneutical proposal I have outlined in this book is *limited* to statements about the natural world in the Bible. Holy Scripture has many different types of literary genres, and therefore every passage requires a specific set of interpretive principles in order to be understood. The series of twenty-two hermeneutical principles were designed specifically for biblical passages dealing with the structure, operation, and origin of the universe and life. In other words, the non-concordist hermeneutical approach that I am proposing has a restricted application and it will not answer all our questions about biblical interpretation.

Concordism & Anti-Evolutionism

I suspect that the most disturbing idea for many Christians regarding these hermeneutical principles is that scientific concordism fails. Many of us are taught in church and Sunday school to believe that the Bible in some way aligns with the facts discovered by modern science. However, if we read statements about nature in Scripture through ancient eyes and with an ancient mindset, it becomes quite clear that the Bible includes ancient science. Figure 1 is a summary of the biblical evidence found in this book to demonstrate that there is no alignment between Scripture and physical reality. In addition, we noted that Jesus employed ancient geography, ancient astronomy, and ancient biology in his teaching ministry (Fig. 22-1, p. 188). If the Lord used the science-of-the-day, then we as Christians must acknowledge this fact in our interpretation of the Word of God.

Another troubling notion in this book for several Bible-believing Christians is that the anti-evolutionary views of young earth creation and progressive creation are based on a concordist hermeneutic, and ultimately rooted in an ancient biology. The creation of living organisms "after their/its kinds" as stated ten times in Genesis 1 reflects an ancient taxonomy that plants and animals are immutable and do not change. In thinking about origins, ancient people like the biblical writers saw that a creature always gave birth to the same type of creature, and they retrojected this observation back to the beginning of the world. The inspired authors of Scripture then came to the reasonable conclusion that the Creator had created each kind of plant and animal *de novo*—quickly and completely.

However, the *de novo* creation of living organisms in Genesis 1 and 2 is an ancient science of origins. This divine creative action is no different than the *de novo* creation of the firmament on the second day of creation, or the *de novo* creation and placement of the sun, moon, and stars in this heavenly dome on the fourth creation day. But everyone knows that there is no solid structure embedded with these heavenly bodies above us. Young earth creation and progressive creation fail to recognize and respect that God's creative action in Genesis 1 and 2 is accommodated/filtered through an ancient scientific understanding of origins. As a consequence, the concordist hermeneutic of the anti-evolutionary positions forces them to accept a God-of-the-gaps view of how the world was made.

ANCIENT GEOGRAPHY	PHYSICAL REALITY
Earth is flat & immovable	No
Earth has foundations set in waters of the deep	No
Earth is circular & has ends	No
Circumferential sea is flat & surrounds circular earth	No
Circumferential sea bordered by ends of heaven	No
Underworld exists below surface of earth	No
Earth begins as formless, empty, dark & watery state	No
De Novo creation of outer border of circumferential sea	No
De Novo creation of waters below (seas) in one place	No
De Novo creation of dry land rising out of seas	No

ANCIENT ASTRONOMY

Firmament is a solid domed structure above earth	No
Firmament has ends at the horizon	No
Firmament is set on foundations	No
Heavenly waters above held up by firmament	No
Divine dwelling set in heavenly body of water	No
Sun, moon & stars placed in firmament	No
Sun moves across dome of firmament daily	No
Stars dislodge from firmament & fall to earth	No
De Novo creation of firmament	No
De Novo creation of heavenly waters above	No
De Novo creation of sun, moon & stars	No

ANCIENT BIOLOGY

Mustard seed is smallest of all seeds	No
Plant growth is controlled only by the soil	No
Seeds die before germination	No
Bats are birds & rabbits are ruminants (chew cud)	No
Males have reproductive seed	No
Females do not have reproductive seed	No
Womb of females like a field for seed of males	No
Infertility caused only by females	No
Plants are immutable (unchanging)	No
Animals are immutable (unchanging)	No
Humans are immutable (unchanging)	No
Living organisms created "according to their kind"	No
Land animals created by sprouting out of the earth	No
De novo creation of plants	No
De novo creation of sea creatures	No
De novo creation of birds	No
De novo creation of land animals	No
De novo creation of 1st man (Adam) from dust	No
De novo creation of 1st woman (Eve) from side of man	No

Figure 1. The Ancient Science in the Bible & the Failure of Scientific Concordism

Now it is crucial to underline again that the Holy Spirit did not lie in the biblical creation accounts. Lying requires the intent to deceive, and the God of the Bible is not a God of deception. Instead, the Lord accommodated. The Holy Spirit allowed the inspired writers of Genesis 1 and 2 to employ the origins science-of-the-day as a vessel to deliver the central inerrant truth that the God of Scripture is the Creator of plants and animals, including men and women.

In light of our twenty-two hermeneutical principles, we can now answer the question that was asked in the title of the Introduction: Is the Bible a Book of Science? My answer: *The Bible is not a book of science.* In particular, the Word of God does not reveal how God actually created the universe and life.

Free from the chains of scientific concordism, we can now explore the majestic creation without *fear* or *guilt* and even study the Lord's creative process by using modern scientific methods and instruments. We should be grateful that our Creator has gifted us with the intellectual ability to explore his marvellous world. In my own scientific research, reflections of intelligent design in the evolutionary process strike me with overwhelming force. Science leads me to an Intelligent Designer. Freedom from concordism has opened my eyes to see that the Lord has created the cosmos and living organisms through an ordained, sustained, and intelligently designed self-assembling evolutionary process. Yes, I am an evolutionary creationist.

Toward a Non-Concordist Hermeneutical Approach

The most important interpretive concept in developing a non-concordist interpretation for statements about nature in Scripture is the Message-Incident Principle. First and foremost, it directs our attention to the life-changing messages of faith in the Bible. This principle recognizes that these inerrant spiritual truths are delivered by an incidental ancient science that is based on an ancient phenomenological perspective of nature. Therefore, it is critical that we *separate* these eternal transcendent truths from their ancient scientific vessels, and *not conflate* (blend) the two together. Regrettably, the conflation of spiritual messages and ancient sciences in Scripture is a common problem within our churches and Sunday schools. As a result, many Christians believe that statements in Scripture about the physical world and its origins are inerrant. But this is a mistake, because the Bible has an ancient science.

Another problem that often appears within church circles is that Bible-believing Christians eisegetically force their modern twenty-first century science into the Word of God. Remember our thought experiment in Hermeneutical Principle 4 where we pictured the shape of the earth in Genesis 1:2. Most people envision a spherical planet. However, historical criticism leads us away from making this interpretive error, because it reveals that ancient people believed the earth was flat and circular. By studying the literature and archeology of ancient nations surrounding the ancient Hebrews and early Christians, it becomes evident that the Bible includes incidental ancient elements. As George Eldon Ladd states, "The Bible is the Word of God given in the words of men in history."[1] Historical criticism assists in identifying ancient characteristics in Scripture and allows us to separate these incidental vessels from the inerrant spiritual truths that they deliver.

In Hermeneutical Principle 2, we noted that determining the literary genre of a written work is one of the most critical decisions in hermeneutics. If we misidentify the genre of a biblical passage, then we will misinterpret it. To be sure, determining the literary genre of the biblical accounts of origins in Genesis 1-3 is quite challenging, and most Christians throughout history have embraced various concordist interpretations of these chapters. But let me suggest that we need to move toward accepting a non-concordist literary genre.

Simply stated, I believe the literary genre of Genesis 1-3 is an *ancient account of origins*.[2] Notably, it is deeply rooted in an ancient science. These chapters are also made up of two sources. The Priestly creation account in Genesis 1 features ancient poetry and is structured on a pair of parallel panels that directs us to honor the Sabbath. The Jahwist account in Genesis 2-3 is a free-flowing parable-like story with many allegorical characteristics, and it emphasizes our sinfulness. Though the idea of a non-literal and non-historical story of origins in Scripture troubles many Christians, we need to remember that Jesus himself used parables in a third of his teaching. These are stories of events that never actually happened. Therefore, the Lord offers a way for us to believe that spiritual truths can be revealed through non-literal stories, such as those in the biblical creation accounts. In addition, the clear evidence of ancient science in Scripture indicates that the Holy Spirit delivered eternal and transcendent messages of faith without using true scientific facts.

I also believe that the Holy Spirit inspired the Priestly author, the Jahwist author, and the editor (redactor) who placed the P and J accounts of origins together in Genesis 1-3. As a result, the Bible opens with complementary inerrant revelations of a heavenly God that rules over his grand creation, and who at the same time, is an earthly Lord intimately involved in the personal lives of men and women. In this way, the non-concordist literary genre of Genesis 1-3 with its ancient poetry, parable-like story, and ancient science do not reveal *how* the world was created, but *who* created it—the God of the Bible.

The interpretation of the creation accounts in Scripture has significant implications for the church. It is well-known that young people today are leaving the faith in record numbers and at a record speed. According to one survey, science and the issue of origins is one of the reasons for this shocking loss of faith. The study records that 25% of our youth perceive that "Christianity is anti-science," and 23% have "been turned off by the evolution vs. creation debate."[3] In another study, 49% of church-attending teenagers say that "the church seems to reject much of what science tells us about the world."[4] Scientific con-cordism fuels this spiritual tragedy.

I have had the privilege of teaching courses on science and religion to university students for nearly twenty-five years, and I know that most of them yearn for a way to embrace both modern science and conservative Christian faith. The first step in developing a peaceful and God-honoring relationship is discovering a non-concordist hermeneutical approach. Once freed from the assumption that the Bible is a book of science, students experience their faith strengthened by the God-glorifying discoveries of science today, including the science of evolutionary biology. In moving beyond scientific concordism, the Bible becomes for them a source of inerrant spiritual truths about our Creator, the creation, and us. And my students share with me that learning about a complementary relationship between Scripture and science fulfills a part of Jesus' first commandment, "Love the Lord your God . . . with all your mind" (Matt. 22:37). Humbled and blessed by these young men and women, to this I say a hardy Amen!

Christian Positions on the Origin of the Universe & Life

	YOUNG EARTH CREATION Six Day Creation	PROGRESSIVE CREATION Day-Age Creation	EVOLUTIONARY CREATION Theistic Evolution
Intelligent Design	Yes Nature points to a Designer	Yes Nature points to a Designer	Yes Nature points to a Designer
Age of the Universe	Young 6000 years	Old 14 billion years	Old 14 billion years
Evolution of Life	No	No	Yes
God's Activity in the Origin of the Universe & Life	Yes Miraculous events for living organisms & inanimate universe over 6 days	Yes 1. Miraculous events for living organisms 2. Natural processes for inanimate universe	Yes Ordained & sustained natural processes for living organisms & inanimate universe
God's Activity in the Lives of Men & Women	Yes Dramatic & subtle miracles	Yes Dramatic & subtle miracles	Yes Dramatic & subtle miracles
Interpretation of Genesis 1	Accepts Spiritual Truths Strict concordism Creation days = 24 hrs The Bible is used as a book of science	Accepts Spiritual Truths General concordism Creation days = millions of yrs The Bible is used as a book of science	Accepts Spiritual Truths Rejects concordism Recognizes ancient science The Bible is NOT a book of science

APPENDIX 1

APPENDIX 2

The "Waters Above" & Scientific Concordism

The formation of the "waters above" as described on the second day of creation in Genesis 1 has led Christians to a number of different concordist interpretations. In this appendix we will examine two popular interpretations of this heavenly body of water: (1) the water canopy theory, and (2) the water vapour theory

Water Canopy Theory

Young earth creationists claim that the "waters above" created on the second day of creation in Genesis 1:6-7 was a water canopy that at one time was above the earth and completely surrounded it. They also assert that this canopy collapsed and contributed to the waters covering the earth during Noah's worldwide flood as described in Genesis 6-9.

In the book that ushered in the modern young earth creationist movement, *The Genesis Flood: The Biblical Record and Its Scientific Implications*, Henry Morris and John Whitcomb, Jr. explain the canopy theory.

> On the second day of Creation, the waters covering the earth's surface were divided into two great reservoirs—one below the firmament and one above; the firmament being the "expanse" above the earth now corresponding to the troposphere... If we accept the biblical testimony concerning an antediluvian [pre-flood] canopy of waters (Gen. 1:6-8, 7:11, 8:2, 2 Pet 3:5-7), we have an adequate source for the waters of a universal flood.[1]

THE BIBLE & ANCIENT SCIENCE

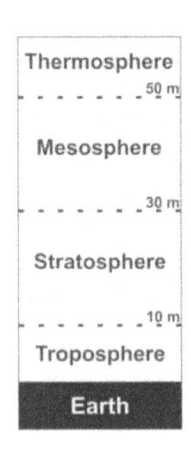

Figure 1. The Canopy Theory & Structure of the Atmosphere.
Note: troposphere surrounding earth not drawn to scale; m = miles

Figure 1 sketches the water canopy theory and shows the modern scientific understanding of the atmosphere. I am sure that you have identified many biblical and scientific problems with this concordist interpretation.

First, Morris and Whitcomb eisegetically assume that the earth is spherical in the Bible. Second, they have a mistaken understanding of the word "firmament." They claim it is an atmospheric "expanse" that corresponds to the troposphere, which is the first ten miles of atmosphere above the earth. But the Hebrew word *rāqîa'* refers to a solid physical dome overhead. Third, on the fourth creation day in Genesis 1:14-17, God places the sun, moon, and stars *in* the firmament. Obviously, this cannot be the troposphere. Finally, Psalm 104:2-3, Psalm 148:3-4, and Jeremiah 10:12-13 all refer to the heavenly sea. These passages were written well *after* Noah's flood. Therefore, the purported collapse of the water canopy claimed by Morris and Whitcomb is not consistent with Scripture.

Water Vapor Theory

Progressive creationists assert that the "waters above" mentioned on the second creation day in Genesis 1:6-7 refer to water vapor in the atmosphere. They contend that this gaseous water also includes the clouds, and that God created the water cycle in this passage.

In his book *The Genesis Question: Scientific Advances and the Accuracy of Genesis*, leading progressive creationist Hugh Ross offers an outline of the water vapor theory.

> On [Creation] Day Two, "God made the expanse and separated the water under the expanse from the water above it" (Genesis 1:7) . . . God's "separation" of the water accurately describes the formation of the troposphere, the atmospheric layer just above the ocean where clouds form and humidity resides, as distinct from the stratosphere, mesosphere, and ionosphere lying above.[2]

Ross claims that the Hebrew word *rāqîaʻ* refers to the "expanse" and defines it as "the atmosphere immediately above the surface of the earth."[3] He then adds that the "water cycle begins" on "day two" of Genesis 1.[4]

Figure 2 is a diagram of the water vapor theory. Once again there are many obvious problems with this concordist interpretation of Scripture and its alignment with modern science.

First, Ross eisegetically assumes that the earth in Scripture is a sphere. Second, the Hebrew word *rāqîaʻ* is not the troposphere, but refers to the firmament, the solid dome of heaven. Third, Ross misreads Scripture. Genesis 1:7 states that the "waters above" are *above* the firmament, and not *in* the firmament. Fourth, if the inspired writer of Genesis 1 had intended to assert that God made water vapor or clouds on the second day of creation, he had three well-known Hebrew words he could have used (*'ēd, nāśî', 'ānān*). But instead he used *mayim* five times in referring to

Figure 2. The Water Vapor Theory. Note: troposphere not drawn to scale

liquid water. Fifth, the Bible states that on the fourth creation day God put the sun, moon, and stars in the *rāqîa'*, and obviously it is not possible to place these heavenly bodies in the troposphere.

Finally, it has long been noted that God does not declare his creative acts on the second day of creation as "good." One would think that in a hot and arid region like ancient Israel, the creation of precipitation and the water cycle would be incredibly good! But the reason this divine declaration does not appear is because day two is not dealing with the hydraulic cycle. It refers to the first phase in the creation of the heavens—the making of the solid dome of the firmament and the sea of waters above it. During the second phase, the heavens are finished on the fourth day of creation with the placement of the sun, moon, and stars in the firmament. It is only upon the completion of the heavens that "God saw it was good" (Gen. 1:18).

To conclude, the water canopy and the water vapor theories demonstrate the problem with attempts to align Scripture with science. The Word of God features an ancient science, and as a consequence, scientific concordism is not possible. The "waters above" on the second day of creation refers to a heavenly sea being held up by a solid firmament. Concordist interpretations of these waters are eisegetical and unbiblical.

APPENDIX 3

Do Isaiah 40:22 & Job 26:7 Refer to a Spherical Earth?

In light of our examination of the 3-tier universe in Hermeneutical Principles 15 and 16, we can now deal with two well-known verses that many Christians use in an attempt to prove that the Bible reveals the modern scientific idea that the earth is a sphere.

Scientific concordists often present Isaiah 40:22 simply as, "He [God] sits enthroned above the circle of the earth," and follow this verse with Job 26:7, "He [God] spreads out the northern skies over empty space; and he suspends the earth over nothing." According to a concordist interpretation, the prophet Isaiah is referring to the outline of planet earth viewed from outer space, and Job presents the earth being suspended by the forces of gravity.

However, these two examples are classic biblical proof-texts. These verses are ripped out of their ancient scientific context, and then manipulated by eisegetically reading into them modern concepts of science that were never intended by the human author or the Holy Spirit.

As we noted earlier, the "circle of the earth" in Isaiah 40:22 refers to the circumferential shore of a flat circular earth (Figs. 14-4, 14-6, 15-1, 15-2; pp. 103, 105, 109, 114). The entire verse reads, "He [God] sits enthroned above the circle of the earth, and its people are like grasshoppers. He stretches out the heavens like a canopy, and spreads them out like a tent to live in." This passage clearly reflects an ancient understanding of the structure of the universe with the domed canopy of heaven above and the flat floor of earth below. The tent analogy for the structure of the world is also used in Psalms 19:4-5 and 104:2-3.

In addition, Isaiah 40:22 is consistent with the ancient geography found in other passages in this biblical book. The prophet Isaiah mentions the "foundation/s of the earth" four times (24:18; 48:13; 51:13, 16). For example, Isaiah 48:13 states, "My [God's] own hand laid the foundations of the earth." The phrase the "ends of the earth" appears nine times (5:26; 40:28; 41:5, 9; 42:10; 48:20; 49:6; 52:10; 62:11). Isaiah 40:28 claims that God is "the Creator of the ends of the earth." And Isaiah refers to the underworld using the Hebrew word *she'ōl* six times (5:14; 14:9, 15; 28:15. 18; 57:9). In a judgment against the King of Babylon, Isaiah 14:9 declares, "The realm of the dead below [*she'ōl*] is all astir to meet you at your coming; it rouses the spirits of the departed to greet you."

The Book of Isaiah also features an ancient astronomy. Isaiah 13:5 uses the phrase the "ends of the heavens" to mean the end of the firmament at the horizon. In describing the final days of the world, Isaiah 34:4 states, "All the stars in the sky will be dissolved and the heavens rolled up like a scroll; all the starry host will fall like withered leaves from the vine, like shriveled figs from the fig tree." The rolling up of the heavens clearly indicates that it is a physical structure (i.e., the firmament), and not immaterial outer space. Similarly, for stars to fall to earth would mean that they are merely tiny specks attached to the firmament.

The common scientific concordist interpretation of Job 26:7 asserts that the earth is suspended in outer space. Of course, this approach eisegetically assumes that the earth is a spherical planet. But Job 11:8 reflects a 3-tier universe. "They [mysteries of God] are higher that the heavens above . . . they are deeper than the depths below [*she'ōl*; the underworld]." Similar to Isaiah 40:22, Job 28:24 points to this ancient understanding of the structure of the world. "For he [God] views the ends of the earth and sees everything under the heavens." In other words, God is overhead in heaven and he observes the shoreline of the circular flat earth and everything under the inverted bowl-shaped firmament.

An ancient geography also appears in the Book of Job. The foundations or pillars of the earth are mentioned twice. In Job 38:4, God asks

Job, "Where were you when I laid the earth's foundation?" Job 9:6 states, "He [God] shakes the earth from its place and makes its pillars tremble." The expression the "ends of the earth" is found two times. In Job 28:24 as noted above, and in Job 37:3, "He [God] unleashes his lightning beneath the whole heaven and sends it to the ends of the earth." The creation of the outer boundary of the circumferential sea appears in Job 26:10. "He [God] marks out the horizon on the face of the waters for a boundary between light and darkness." And Job mentions the underworld eight times, using the Hebrew word *she'ōl* (7:9,11:8, 14:13, 17:13, 16; 21:13, 24:19, 26:6). Job 17:16 asks, "Will it [my hope] go down to the gates of death [*she'ōl*]?

This biblical book also includes an ancient astronomy. In Job 26:11, Job claims, "The pillars of the heavens quake, aghast at his [God's] rebuke." The notion that the heavens had foundations made sense, because the inspired author believed in the firmament, as seen in Job 37:18, "Can you join him [God] in spreading out the skies [*sheḥāqîm*] hard as a mirror of cast bronze?" As we have noted, *sheḥāqîm* is another Hebrew word for the firmament. Reference to the firmament also appears in the term the "vaulted heavens" in Job 22:14. And the location of God's dwelling in the 3-tier universe is reflected in Job 36:29. "Who can understand . . . how he [God] thunders from his pavilion?" In other words, the Lord lives just overhead in the upper heavens where the rumble of thunder arises.

It must be added that the word translated as "suspends" in Job 26:7 is the Hebrew verb *tālāh*. While the word "suspend" in English can carry the nuance of suspending or floating an object in air without attaching to anything, in the Bible *tālāh* appears in the context of hanging an object on something. For example, utensils are hung on a peg (Isa. 22:23-24), shields and helmets on a wall (Ezek. 27:10), and a harp on a tree (Ps. 137:2). Notably, 20 of the 28 times *tālāh* is used in Scripture refer to the execution of a person by hanging them from a tree or gallows (e.g., Gen. 40:22; Deut. 21:22; Est. 7:10). This Hebrew verb never means to suspend something in empty space.

THE BIBLE & ANCIENT SCIENCE

To conclude, Isaiah 40:22 and Job 26:7 must be understood within their ancient scientific context. The circle of the earth in Isaiah does not refer to the outline of planet earth from outer space, but to a circular flat earth surrounded by a flat circumferential sea. The earth being suspended over nothing in Job does not refer to our planet hovering in empty space. Rather, it simply means that the flat earth is not hung from anything in the universe, such as the firmament overhead. Scientific concordist interpretations of Isaiah 40:22 and Job 26:7 are eisegetical and unbiblical.

Notes

Introduction: Is the Bible a Book of Science?
[1] Dennis R. Venema and Scot McKnight, *Adam and the Genome: Reading Scripture after Genetic Science* (Grand Rapids, MI: Brazos Press, 2017), 104-5.
[2] Ibid., 171.
[3] Henry M. Morris, *Many Infallible Proofs: Practical and Useful Evidences of Christianity* (San Diego, CA: Creation-Life Publishers, 1980), 229. My italics.
[4] David Frost, *Billy Graham: Personal Thoughts of a Public Man. 30 Years of Conversations with David Frost* (Colorado Springs, CO: Chariot Victor, 1997), 73. My italics
[5] Johannes Kepler to Herwart von Hohen, Letter dated 9/10 April 1599, in Carola Baumgardt and Jamie Callan, *Johannes Kepler Life and Letters* (London, UK: V. Gollancz, 1952), 50.

Principle 1: Literalism
[1] Martin Luther, *Luther's Works. Lectures on Genesis: Chapters 1–5*. Jaroslav Pelikan, ed. (Saint Louis, MO: Concordia Publishing House, 1958), 5.
[2] No Author, "Six in Ten Take Bible Stories Literally, But Don't Blame Jews for Death of Jesus." No pages. Survey conducted 6–10 February 2004 with a random sample of 1011 adults by ICR-International Communications Research Media, PA. Accessed March 7, 2015: http://abcnews.go.com/images/pdf/947a1ViewsoftheBible.pdf

Principle 2: Literary Genre
[1] Benedict T. Viviano refers to Matthew 5:21-48 as "six hypertheses." Instead of the common genre designation of "antitheses," he argues that "Jesus seems to go beyond OT teaching by deepening and radicalizing it." Benedict T. Viviano, "The Gospel according to Matthew" in R.E. Brown, J.A. Fitzmyer, and R.E. Murphy, eds. *The New Jerome Biblical Commentary*, 2nd ed. (New York, NY: Geoffrey Chapman, 1990), 641. Yet for me, a problem remains: I am doubtful that Jesus intended these commands to be literal, like plucking out eyes and cutting off hands.
[2] For two excellent introductions on this topic, see Craig D. Allert, *Early Christian Readings of Genesis One: Patristic Exegesis and Literal Interpretation* (Downers Grove, IL: InterVarsity Academic, 2018); Peter C. Bouteneff, *Beginnings: Ancient Christian Readings of the Biblical Creation Narratives* (Grand Rapids, MI: Baker Academic, 2008).
[3] Ken Ham, *The Lie: Evolution. Genesis—The Key to Defending Your Faith* (Green Forest, AR: Master Books, 2012), 10.
[4] For an excellent introduction to the historicity of the Old Testament, see William G. Dever, *What Did the Biblical Writers Know and When Did They Know It? What Archeology Can Tell Us about the Reality of Ancient Israel* (Grand Rapids, MI: Eerdmans, 2001).
[5] For an excellent introduction to the historicity of the New Testament, see Richard Bauckham, *Jesus and the Eyewitnesses: The Gospels as Eyewitness Testimony* (Grand Rapids, MI: Eerdmans, 2006).

Principle 3: Scientific Concordism & Spiritual Correspondence
[1] Henry M. Morris, *Many Infallible Proofs: Practical and Useful Evidences of Christianity* (San Diego, CA: Creation-Life Publishers, 1980), 229. My italics.

[2] David Frost, *Billy Graham: Personal Thoughts of a Public Man. 30 Years of Conversations with David Frost* (Colorado Springs, CO: Chariot Victor, 1997), 73-74. My italics.

[3] No Author, "Six in Ten Take Bible Stories Literally, But Don't Blame Jews for Death of Jesus." No pages. Survey conducted 6–10 February 2004 with a random sample of 1011 adults by ICR-International Communications Research Media, PA. Accessed March 7, 2015: http://abcnews.go.com/images/pdf/947a1ViewsoftheBible.pdf

[4] In some of my earlier publications I used the category "spiritual concordism," but was never comfortable with this term.

[5] Henry M. Morris, "Foreword" in John D. Morris, *The Young Earth* (Colorado Springs, CO: Creation-Life Publishers, 1994), 5.

[6] Some might complain that the term "ancient science" is anachronistic. However, ancient people were enthralled by nature and they certainly attempted to understand its origin, operation, and structure. For a superb study demonstrating the remarkable knowledge of nature in the ancient world, see David C. Lindberg, *The Beginnings of Western Science: The European Scientific Tradition in Philosophical, Religious, and Institutional Context, 600 B.C. to A.D. 1450*, Chicago, IL: University of Chicago Press (1992). Note Lindberg's use of the terms "Science" and "Scientific" in the title of his book and the dates of his extensive study. Clearly, this historian of science does not view these terms as anachronistic.

Principle 4: Eisegesis vs. Exegesis

[1] The translations of Greek and Hebrew words in this book are common renditions from standard lexicons. These are not cited in each case and include: Walter Bauer, *A Greek-English Lexicon of the New Testament and other Early Christian Literature*, William F. Arndt and F. Wilbur Gingrich, eds. (Chicago, IL: University of Chicago Press, 1958); Henry George Liddell, *A Greek-English Lexicon*, revised by Henry Stuart Jones (Oxford, UK: Oxford University Press, 1996); Francis Brown, S.R. Driver and C.A. Briggs, *Hebrew and English Lexicon of the Old Testament* (Oxford, UK: Clarendon Press, 1951); David J.A. Clines, ed., *The Dictionary of Classical Hebrew*, 8 vols. (Sheffield, UK: Sheffield Academic Press.2010); Ludwig Koehler and Walter Baumgartern, *The Hebrew and Aramaic Lexicon of the Old Testament*, 5 vols. (New York, NY: Brill Academic, 1994).

[2] Martin Luther, *Luther's Works. Lectures on Genesis*, ed. Jaroslav Pelikan 79 vols. (St. Louis: Concordia Publishing House, 1958), I:42.

[3] Ibid., LIV: 358-9.

Principle 5: Ancient & Modern Phenomenological Perspectives

[1] I am grateful to John H. Walton for this hermeneutical insight. See his *The Lost World of Genesis One: Ancient Cosmology and the Origins Debate* (Downers Grove, IL: IVP Academic, 2009), 9.

[2] Of course, it was Nicholas Copernicus who introduced this idea in the sixteenth century, but Galileo certainly popularized it.

[3] For an excellent overview of ancient creation accounts throughout the world, see David A. Leeming, *Creation Myths of the World: An Encyclopedia*. 2 vols. 2nd ed. (Santa Barbara, CA: ABC-CLIO, LCC, 2010).

Principle 6: The Message-Incident Principle

[1] We will examine some Mesopotamian creation accounts in more detail in Hermeneutical Principle 17.

Principle 7: Biblical Accommodation
[1] John D. Miller, Eugenie Scott, and Shinji Okamoto, "Public Acceptance of Evolution Science," *Science* 313 (11 August 2006), 765-6.

Principle 8: Authorial Intentionality: Divine & Human
[1] Henry M. Morris, *Many Infallible Proofs* (San Diego, CA: Creation Life Publishers, 1980), 230.
[2] Hugh Ross, *The Genesis Question: Scientific Advances and the Accuracy of Genesis* (Colorado Springs, CO: NavPress, 1998), 17 and 19.

Principle 9: Biblical Sufficiency & Human Proficiency
[1] It is possible that Paul could have held a geocentric understanding of the cosmos (see Figures 9-1 and 9-2) with the "underworld" in the core of the earth. Nevertheless, my point remains in that he accepted an ancient geography. It is interesting to note that in the fifth century, debate existed within the Church regarding whether the structure of the world was 3-tiered or geocentric. See St. Augustine, *Literal Meaning of Genesis*, John Hammond Taylor, trans. 2 vols. (New York: Newman Press, 1982), 1:58–59.
[2] For the sake of simplicity and illustration, I have not included in Figure 9-2 the ancient astronomical concepts of epicycles, deferrents, equants, etc. For a helpful introduction, see Charles E. Hummel, *The Galileo Connection: Resolving Conflicts between Science and the Bible* (Downers Grove, IL: InterVarsity Press, 1986).

Principle 10: Modern Science & Paraphrase Biblical Translation
[1] Eugene H. Peterson, *The Message New Testament; The New Testament in Contemporary Language* (Colorado Springs, CO: NavPress, 1993), 82.
[2] Ibid. My italics.
[3] Ibid., 7.
[4] Ibid.
[5] Ibid., 6-7.
[6] Ibid., 7.
[7] Ninety-eight percent of American scientists accept that "humans and other living things have evolved over time." No Author, "Elaborating on the Views of AAAS Scientists, Issue by Issue" Pew Research Center: Science and Society (23 July 2015). Accessed February 7, 2019: http://www.pewresearch.org/science/2015/07/23/elaborating-on-the-views-of-aaas-scientists-issue-by-issue/.

Principle 11: Textual Criticism
[1] Alfred Marshall, *The New International Version Interlinear Greek-English New Testament* (Grand Rapids, MI: Zondervan, 1976), 153.
[2] K. Aland, M. Black, C. Martini, B. Metzger and A. Wikgren, *The Greek New Testament*, 3rd ed. (West Germany: United Bible Societies, 1983), 136.
[3] This is the problem with the KJV translation. It assumes that the first two verses of Scripture present a sequential order of events.
[4] Claus Westermann, *Genesis 1–11: A Commentary*, John J. Scullion trans. (Minneapolis, MN: Augsburg Publishing House, 1987), 78, 93–102; Gordon J. Wenham, *Genesis 1–15* (Waco: Word Books, 1987), 11-17.

Principle 12: Implicit Scientific Concepts

[1] In a recent scientific paper on dentition development, I refer to an "odontogenic field" in which teeth initiate like plants sown within the boundaries of a farmer's field. Denis O. Lamoureux, Aaron R. H. LeBlanc, and Michael W. Caldwell. "Tooth germ initiation patterns in a developing dentition: An *in vivo* study of *Xenopus laevis* tadpoles." *Journal of Morphology* 279:5 (May 2018), 616-625; DOI: 10.1002/jmor.20797.

[2] I am using my own translations in this paragraph because the 2011 NIV Bible loses the thrust of the original Hebrew.

[3] Needham notes, "We know that the conception of the female sex as playing the part of farm-land, i.e., of woman as a field in which grain was sown, was widespread in antiquity. Joseph Needham, *A History of Embryology* (New York, NY: Arno Press, 1975), 44.

[4] St. Augustine offers evidence of this ancient understanding of reproductive biology in Hebrews 7:10. He claims, "Levi was there [in Abraham's loins] according to the seminal [seed] principle by which he was destined to enter his mother on the occasion of carnal union." St. Augustine, *The Literal Meaning of Genesis*, John H. Taylor, trans., 2 vols. (New York, NY: Newman Press, 1982 [415]), II:123.

[5] Needham observers, "By 1720 the theory of preformatism was thoroughly established, not only on the erroneous grounds put forward by Malpighi and Swammerdam, but on the experiments of Andry, Dalenpatius and Gautier, who all asserted they had *seen exceedingly minute forms of men, with arms, heads, and legs complete, inside the spermatozoa* under the microscope." Needham, *History of Embryology*, 205. My italics.

[6] N. Hartsoecker, *Essai de Dioptrique* (1695). *Preformation* [Online image depicting preformation (fully formed human within sperm]. Retrieved December 23, 2016 from: https://en.wikipedia.org/wiki/File:Preformation.GIF This image is considered public domain.

[7] Hugh Ross, *Navigating Genesis: A Scientist's Journey through Genesis 1-11* (Covina, CA: RTB Press, 2014), 29.

[8] Ibid., 29. Redrawn from Figure 3.

[9] Hugh N. Ross, *Creation and Time* (Colorado, CO: NavPress, 1994), 153. Christian tradition suggests that Moses is the author of the Book of Genesis.

Principle 13: Scope of Cognitive Competence

[1] I thank botanist Jack Maze for this statistic. Also see Oren L. Justice and Louis N. Bass, *Principles and Practices of Seed Storage* (Washington, DC: U.S. Government Printing Office, 1978), 203-204. I am very grateful to horticulturalist Keith Furman for this citation and his assistance.

[2] Needham notes, "We know that the conception of the female sex as playing the part of farm-land, i.e., of woman as a field in which grain was sown, was widespread in antiquity." Needham, *History of Embryology*, 44.

[3] Susan C. Stuart, "Male Infertility." No pages. Accessed July 26, 2007. Online: http://www.thedoctorwillseeyounow.com/articles/other/malein_29/#back2

Principle 14: Historical Criticism

[1] For example, an excellent resource is Kenton L. Sparks, *Ancient Texts for the Study of the Hebrew Bible* (Peabody, MA: Hendrickson Publishers, 2005).

[2] Aeschylus, *Aeschylus I: Oresteia*, Richard Lattimore, trans. (Chicago, IL: University Press, 1953), 158.

[3] George Eldon Ladd, *The New Testament and Criticism* (Grand Rapids, MI: Eerdmans, 1967), 12.

[4] John H. Walton, *The Lost World of Adam and Eve: Genesis 2-3 and the Human Origins Debate* (Downers Grove, IL: IVP Academic, 2015), 19. My italics.

[5] John H. Walton, *The Lost World of Genesis One: Ancient Cosmology and the Origins Debate* (Downer Groove, IL: IVP Academic, 2009), 19. My italics.

[6] Ibid., 106. My italics.

[7] Othmar Keel, *The Symbolism of the Biblical World* (New York, NY: Seabury Press, 1978), 36. Permission to copy image kindly granted by Penn State University Press.

[8] *Ibid.*, 38. Permission to copy image kindly granted by Penn State University Press.

[9] J.H. Breasted, *Ancient Records of Egypt* 5 vols. (New York, NY: Russell and Russell, 1962), II:89 no. 220. Similarly, another inscription states of this queen, "Her fame has encompassed the 'Great Circle' (Okeanos)." *Ibid.*, II:137 no. 325. I am indebted to Paul Seely for these references. In my opinion, his three papers on the ancient science in Scripture and the ancient world are some of the best summaries on this topic. Seely, Paul H. "The Firmament and the Water Above. Part I: The Meaning of *raqia'* in Gen 1:6-8" *Westminster Theological Journal* 53 (1991), 227-240; "The Firmament and the Water Above. Part II: The Meaning of 'The Water above the Firmament' in Gen 1:6-8" *Westminster Theological Journal* 54 (1992), 31-46; "The Geographical Meaning of 'Earth' and 'Seas' in Genesis 1:10" *Westminster Theological Journal* 59 (1997), 231-255.

[10] James B. Pritchard, *Ancient Near Eastern Texts Relating to the Old Testament*, 3rd ed. (Princeton, NJ: University Press, 1967), 374.

[11] Breasted, *Ancient Records of Egypt*, IV:38, no. 64.

[12] Ibid., IV:163, no. 308.

[13] Keel, *Symbolism*, 174. Permission to copy image kindly granted by Penn State University Press. Information of the Shamash Tablet from British Museum. Retrieved January 23, 2019 from: http://www.britishmuseum.org/research/collection_online/collection_object_details.aspx?objectId=282224&partId=1.

[14] Wayne Horowitz, *Cosmic Geography* (Winona Lake, IN: Eisenbrauns, 2011), 262-263.

[15] Diagram based on: (1) Map of the World [Online image of a Babylonian World Map, British Museum Image Number: 92687]. Retrieved January 23, 2019 from: http://www.britishmuseum.org/research/collection_online/collection_object_details/collection_image_gallery.aspx?partid=1&assetid=32436001&objectid=362000. Used with permission of The British Museum under a Creative Commons Attribution-NonCommercial-ShareAlike 4.0 International license. © Trustees of the British Museum, (2) Horowitz, *Cosmic Geography*, 21.

[16] *Ibid.*, 30-32.

[17] See the subsection entitled "The Ends & Pillars/Foundations of the Heavens" in Hermeneutical Principle 16.

[18] James B. Pritchard, ed. *Ancient Near Eastern Texts Relating to the Old Testament*, 3rd ed. (Princeton: Princeton University Press, 1969), 69.

[19] Richard J. Clifford, *Creation Accounts in the Ancient Near East and in the Bible* (Washington, DC: Catholic Biblical Association, 1994), 63.

Principle 15: The 3-Tier Universe: Ancient Geography

[1] Recently there has been discussion regarding the identity of the levels in the 3-tier universe. Greenwood contends the three tiers are the heaven, the earth, and the water underneath the earth (the deep). Kyle R.Greenwood, *Scripture and Cosmology: Reading the Bible between the Ancient World and Modern Science* (Downers Grove, IL: IVP Academic, 2015), 71. However, Driggers asserts that viewing the waters below the earth as a third tier in the universe is not accurate. He argues that "this was more extensive than a mere 'tier,' since the cosmic waters existed below and around the earth, as well as above the firmament." Ira Brent Driggers, "New Testament Appropriations of Genesis 1-2" in Kyle R. Greenwood, ed. *Since the Beginning: Interpreting Genesis 1 and 2 through the Ages* (Grand Rapids, MI: Baker Academic, 2018), 56. See Figure 15-1 to picture the argument by Driggers. I will employ the traditional view of the 3-tier universe which is based on three levels of habitation: (1) the heaven above with God and angels, (2) the earth in the middle with living humans, and (3) the underworld below with the souls of dead humans and demon spirits. See the classic study, Luis I.J. Stadelmann, *The Hebrew Conception of the World*, Analecta Biblica 39 (Rome, Italy: Pontifical Biblical Institute, 1970).

[2] Richard J. Clifford, Creation Accounts in the Ancient Near East and in the Bible (Washington, DC: Catholic Biblical Association, 1994), 63.

[3] J.H. Breasted, *Ancient Records of Egypt* 5 vols. (New York, NY: Russell and Russell, 1962), IV:163, no. 308.

[4] Ibid., IV:38, no. 64.

Principle 16: The 3-Tier Universe: Ancient Astronomy

[1] To emphasize the thrust of the original Hebrew words (which are explained in the next paragraph), I have replaced the term "vault" in the 2011 NIV with "firmament," and "sky" with "heaven." Similarly in Psalm 19:1, I have exchanged "sky" with "firmament." In addition, I have replaced the pronoun "it" in Genesis 1:7 with the word "firmament" because this noun appears in the original Hebrew text.

[2] Wayne Horowitz, *Cosmic Geography* (Winona Lake, IN: Eisenbrauns, 2011), 262-263.

[3] Denis O. Lamoureux, Aaron R. H. LeBlanc, and Michael W. Caldwell, "Tooth germ initiation patterns in a developing dentition: An *in vivo* study of *Xenopus laevis* tadpoles." *Journal of Morphology* 279:5 (May 2018), 616-625; DOI: 10.1002/jmor.20797. See also "Models in Science" in Ian G. Barbour, *Religion and Science: Historical and Contemporary Issues* (New York, NY: Harper San Francisco, 1997), 115-119.

Principle 17: Accommodation in God's Creative Action in Origins

[1] Sparks, *Ancient Texts*, 311. This account is written in both Sumerian and Babylonian (Akkadian).

[2] Richard J. Clifford, *Creation Accounts in the Ancient Near East and in the Bible* (Washington, DC: Catholic Biblical Association, 1994), 63-64.

[3] For an excellent study on the Image of God, see J. Richard Middleton, *The Liberating Image: The* Imago Dei *in Genesis 1* (Grand Rapids, MI: Brazos Press, 2005).

[4] Clifford, *Creation Accounts*, 83.

[5] James B. Pritchard, ed. *Ancient Near Eastern Texts Relating to the Old Testament*, 3rd ed. (Princeton: Princeton University Press, 1969), 60-61.

[6] See pages 80-81 for the discussion on creation-out-of-nothing in the Bible.
[7] Pritchard, *Texts*, 69.
[8] Wayne Horowitz, *Mesopotamian Cosmic Geography* (Winona Lake, IL: Eisenbrauns 1998), 112.
[9] Pritchard, *Ancient Near Eastern Texts*, 71.
[10] Horowitz, *Cosmic Geography*, 118.
[11] Stephanie Dalley, *Myths from Mesopotamia: Creation, the Flood, Gilgamesh, and Others* (Oxford, UK: University Press, 1989), 260.
[12] Pritchard, *Ancient Near Eastern Texts*, 70.
[13] Ibid., 71.
[14] Ibid., 69.
[15] Ibid.
[16] Dalley, *Myths from Mesopotamia*, 261.
[17] Ibid.

Principle 18: *De Novo* Creation of Living Organisms: Ancient Biology

[1] Richard J. Clifford, *Creation Accounts in the Ancient Near East and in the Bible* (Washington, DC: Catholic Biblical Association, 1994), 30. This account is also known as "KAR 4."
[2] Ibid.
[3] Walter Beyerlin, ed., *Near Eastern Religious Texts Relating to the Old Testament* (Philadelphia: Westminster Press, 1978), 75.
[4] Clifford, *Creation*, 74. Also Stephanie Dailey, *Myths from Mesopotamia: Creation, The Flood, Gilgamesh, and Others* (Oxford, UK: University Press, 2000), 14-17.
[5] Clifford, *Creation*, 75.
[6] Ibid., 48–49.
[7] Ibid., 105, 107.
[8] Francois Daumas, Mammisis de Dendera, Khnum and Isis (2014). Retrieved April 25, 2017 from: https://commons.wikimedia.org/wiki/File:Chnum-ihy-isis.jpg. This file is licensed under the Creative Commons Attribution-Share Alike 3.0 Unported, 2.5 Generic, 2.0 Generic and 1.0 Generic license.
[9] George Eldon Ladd, *The New Testament and Criticism* (Grand Rapids, MI: Eerdmans, 1967), 12.
[10] John H. Walton, *The Lost World of Adam and Eve: Genesis 2-3 and the Human Origins Debate* (Downers Grove, IL: IVP Academic, 2015), 19. My italics.
[11] Diagram based on Nancy Morvillo, *Science and Religion: Understanding the Issues* (Chichester, UK: Wiley-Blackwell, 2010), p. 57.
[12] Martin Luther, *Luther's Works. Lectures on Genesis: Chapters* 1–5. Jaroslav Pelikan, ed. (Saint Louis: Concordia Publishing House, 1958), 30. Capital letters original.
[13] No Author, "Elaborating on the Views of AAAS Scientists, Issue by Issue" Pew Research Center: Science and Society (23 July 2015). Accessed February 7, 2019: http://www.pewresearch.org/science/2015/07/23/elaborating-on-the-views-of-aaas-scientists-issue-by-issue/.

Principle 19: Does Conservative Christianity Require Scientific Concordism?

[1] St. Augustine, *The Literal Meaning of Genesis*, John Hammond, trans. 2 vols. (New York, NY: Newman Press, 1982 [415]), I:58. Regrettably, the translation of the Latin word "*Litteram*" as "Literal" in the title of this book is somewhat misleading for modern audi-

ences, because Augustine was definitely not a literalist in the sense he accepted a literal reading of Genesis 1 similar to young earth creationists. A more accurate term would be "literary."

[2] Ibid., I:59.
[3] Ibid., I:60.
[4] Ibid., I:59.
[5] See St. Augustine, *City of God* (16.9); Gerald G. Walsh, Demetrius B. Zema, Grace Monahan, and Daniel J. Honan, translators (New York, NY: Doubleday Image Book, 1958), 367.
[6] Augustine, *Literal Meaning*, I:60.
[7] Ibid.
[8] Ibid.
[9] Ibid., 61.
[10] Ibid.
[11] Martin Luther, *Luther's Works: Lectures on Genesis*, Chapters 1-5, J. Pelikan, ed. (St. Louis, MO: Concordia Publishing House, 1958), 35.
[12] Ibid., 24.
[13] Ibid., 42.
[14] Ibid., 42-43.
[15] Ibid., 30.
[16] John Calvin, *Commentary on Genesis*, 2 vols (Grand Rapids, MI: Christian Classics Ethereal Library, 2007 [1554]), I: 25-26. Accessed 24 March 2012 at http://www.ccel.org/ccel/calvin/calcom01.pdf.
[17] Ibid., 114. John Calvin, *Commentary on Psalms*, 2 vols (Grand Rapids, MI: Christian Classics Ethereal Library, 2007 [1557]), I: 314. Accessed 24 March 2012 at https://www.ccel.org/ccel/calvin/calcom08.pdf. See also Davis A. Young, *John Calvin and the Natural World* (Lanham, MD: University Press of America, 2007), 31- 32.
[18] Calvin, *Commentary on Genesis*, I:41.
[19] Ibid.
[20] Ibid., I:42.
[21] Robert White, "Calvin and Copernicus: The Problem Reconsidered" *Calvin Theological Journal* 15 (1980), 236, my italics. I am grateful to David Haitel for this reference.
[22] Luther, *Lectures on Genesis*, 5
[23] Ibid., 3.
[24] John Calvin, *The Institutes of the Christian Religion*, Henry Beveridge, translator (Grand Rapids, MI: Christian Classics Ethereal Library, 1836 [1536]), 143.
[25] No Author, "Six in Ten Take Bible Stories Literally, But Don't Blame Jews for Death of Jesus." No pages. Survey conducted 6–10 February 2004 with a random sample of 1011 adults by ICR-International Communications Research Media, PA. Accessed March 7, 2015: http://abcnews.go.com/images/pdf/947a1ViewsoftheBible.pdf.

Principle 20: Literary Criticism

[1] I should qualify that the living creatures are those that move about in the seas and on dry land. I suspect that ancient people did not view plants in the same way as these animals. It is our modern biology that categorizes plants and animals as living organisms.
[2] See **page 202** for the reason the second day of creation is not deemed "good" by the Creator.

³ Lloyd R. Bailey, *Genesis, Creation and Creationism* (New York: Paulist Press, 1993), 157–160.
⁴ See the next hermeneutical principle for this word play in Hebrew.
⁵ My research included the most important books on this topic: Alexander Heidel, *The Babylonian Genesis: The Story of Creation* (and Related Babylonian Creation Stories), James B. Pritchard's *Ancient Near Eastern Texts Relating to the Old Testament*, Stephanie Dalley's *Myths from Mesopotamia: Creation, the Flood, Gilgamesh, and Others*, William Richard J. Clifford's *Creation Accounts in the Ancient Near East and in the Bible*, Kenton L. Sparks, *Ancient Texts for the Study of the Hebrew Bible*, W. Hallo's *The Context of Scripture: Canonical Compositions from the Biblical World*, and David A. Leeming's two volume *Creation Myths of the World: An Encyclopedia*.
⁶ For these papers, see my curriculum vitae at: www.sites.ualberta.ca/~dlamoure/cv.pdf.
⁷ Tobit 8:6 from NRSV.
⁸ Sirach 17:1 and 25:24 from NRSV.
⁹ Wisdom of Solomon 2:23-24 from NRSV.
¹⁰ Much more could be said about hermeneutics in Jewish circles during this period. For example, most Christians are aware of the unusual and even bizarre use of Old Testament passages by New Testament writers. Many times, passages are taken completely out context. For example, Matthew 2:15 refers to Hosea 1:11, "Out of Egypt I called my son." In the original context of Hosea, it is the nation of Israel that is called out of Egypt. But in Matthew this verse is "Christianized" and refers to Jesus. See Paul's application of this approach in Peter Enns, *The Evolution of Adam: What the Bible Does and Doesn't Say about Human Origins* (Grand Rapids, MI: Brazos Press), 103-113.
¹¹ It must be noted that archetypes can be historical individuals, such as in this passage referring to Jesus. Similarly, Abraham is presented as the archetype of the man of faith and those who believe in God (Rom. 4:11).

Principle 21: Source Criticism
¹ For an excellent introduction to the evidence for sources in the Pentateuch, see Richard E. Friedman, *The Bible with Sources Revealed: A New View into the Five Books of Moses* (New York, NY: HarperSanFrancisco, 2003), 1-31.
² This is the translation of the NRSV and 1982 NIV.
³ The Jahwist author usually uses only *Yahweh* (Lord). But uniquely in Genesis 2-3, *'Elōhîm* [God] is placed next to this divine name a total of 20 times. Friedman suggests that the redactor did this in order "to soften the transition from the P creation, which uses only 'God.'" Friedman, *Sources Revealed*, 35.
⁴ For more examples of word play by the J author, see J. Richard Middleton, "Reading Genesis 3 Attentive to Evolution" in W.T. Cavanaugh and J.K. Smith, eds., *Evolution and the Fall* (Grand Rapids, MI: Eerdmans, 2017), 67-97.
⁵ For an excellent study on the Image of God, see J. Richard Middleton, *The Liberating Image: The Imago Dei in Genesis 1* (Grand Rapids, MI: Brazos Press, 2005).

Principle 22: Biblical Inerrancy: Toward an Incarnational Approach
¹ I want to qualify that though the Bible is both the Word of God and the words of men and women, I am not suggesting a hypostatic union of the divine and human aspects of Scripture as understood with the Incarnation. See Kenton L. Sparks, *God's Word in Hu-*

man Words: An Evangelical Appropriation of Critical Biblical Scholarship (Grand Rapids, MI: Baker Academic, 2008), 253-254; Peter Enns, *Inspiration and Incarnation: Evangelicals and the Problem of the Old Testament* (Grand Rapids, MI: Baker Academic, 2005), 18.

[2] Peter Enns terms this view of the Bible a "scriptural Docetism." Enns, *Inspiration and Incarnation*, 18.

Conclusion

[1] George Eldon Ladd, *The New Testament and Criticism* (Grand Rapids, MI: Eerdmans, 1967), 12.

[2] Readers will note that I do not use the term "myth" to describe the literary genre of Genesis 1-3. This is an explosive word in conservative Christian circles. For most people today, the term "myth" means something that is untrue and utterly false. Christians are particularly opposed to myths because they are denounced in the Bible. The apostle Paul warns in 2 Timothy 4:3-4, "For a time will come when men will not put up with sound doctrine. Instead, to suit their own desires, they will gather around them a great number of teachers to say what their itching ears want to hear. They will turn their ears away from the truth and turn aside to myths." The falsity of myths also appears in 1 Timothy 1:4, 4:7, Titus 1:14, and 2 Peter 1:16. Therefore, this word is usually depicted in a very negative light. However, myth is also a literary genre category that refers to an account that conveys the beliefs and values of a community. Defined in this way, Genesis 1-3 could be termed a "creation myth" in that it outlines inerrant spiritual truths of the Hebrews.

[3] Barna Group, "Six Reasons Young Christians Leave Church," 28 Sept 2011. Accessed February 7, 2019: www.barna.org/teens-next-gen-articles/528-six-reasons-young-christians-leave-church.

[4] Barna Group, "Atheism Doubles Among Generation Z," 24 Jan 2018. Accessed March 31, 2019: www.barna.com/research/atheism-doubles-among-generation-z/

Appendix 2: The "Waters Above" & Scientific Concordism

[1] Henry Morris and John Whitcomb, *The Genesis Flood: The Biblical Record and Its Scientific Implications* (Presbyterian & Reformed Press, 1961), 77, 229

[2] Hugh Ross, *The Genesis Question: Scientific Advances and the Accuracy of Genesis* (Colorado Springs, CO: NavPress, 1998), 34. Note the ionosphere begins in the upper mesosphere and extends through the thermosphere to about 620 miles above the earth.

[3] Ibid., 199, 201.

[4] Hugh Ross, *Navigating Genesis: A Scientist's Journey through Genesis 1-11* (Covina, CA: RTB Press, 2014), 41.

CPSIA information can be obtained
at www.ICGtesting.com
Printed in the USA
BVHW091209231120
593961BV00014B/118

9 781951 252052